Natural History Museums
Directions for Growth

Edited by
Paisley S. Cato
Clyde Jones

Texas Tech Unversity Press
1991

This book was set in 10 on 12 Baskerville and printed on acid-free paper that meets the guidelines for permanence and durability of the Committee on Production Guidelines for Book Longevity of the Council on Library Resources. ∞

Manufactured in the United States of America.

Library of Congress Cataloging-in-Publication Data

Natural history museums : directions for growth / edited by Paisley S. Cato, Clyde Jones.
    p. cm.
  Includes bibliographical references.
  ISBN 0-89672-240-6 (cloth)
  1. Natural history museums. I. Cato, Paisley S. II. Jones, Clyde.
QH70.A1N38 1991
508′.074—dc20   90-24266
                                                                          CIP

91 92 93 94 95 96 97 98 99   9 8 7 6 5 4 3 2 1

Texas Tech University Press
Lubbock, Texas 79409-1037 USA

CONTENTS

The Future

INTRODUCTION

Museums are dynamic organizations—sometimes changing slowly, but changing and evolving nonetheless. Efforts to assess and analyze changes benefit both an individual museum and the profession at large. Such efforts are not merely to document where the profession stands at a point in time, but are to provide some insight to issues that should be further questioned, discussed, and addressed. Considering the diversity of museums that exist, their missions, organizational structures, functions, staff, and so on, it is not surprising that analytical efforts require the perspectives of many individuals.

With these thoughts in mind, a symposium entitled "New Directions and Professional Standards for Natural History Museums" was organized for the 1988 joint meeting of the Mountain-Plains Museums Association and the Midwest Museums Conference in Kansas City. Speakers were invited to submit papers dealing with topics they felt were critical issues for natural history museums. The papers in this volume are based on 18 of the 22 presentations given at the symposium and reflect the diversity of concerns held by museum professionals.

The issues facing natural history museums range from the general to the specific. Several authors considered the role and mission of natural history museums. Humphrey described a model for the philosophy of a university natural history museum, taking into consideration the unique constraints and opportunities offered by a university setting. Laerm and Edwards discussed the results and implications of a survey of state museums of natural history. The role museums might fulfill as part of a state's mandate to manage wildlife was presented by Shropshire and Shropshire. Three authors summarized critical historical perspectives in an effort to question where natural history museums are headed (Porter, Denton, and Lintz), and Choate provided very practical, useful information concerning the availability of foundation funding.

Natural history museums constantly face the problems of preserving and managing their collections. However, during the last 10 years, a more analytical approach has developed in the area of collections care. Traditional methods have been

1

questioned, and there is a recognized need for analyzing routine collection management policies and practices. Rose emphasized the need to address preservation concerns while maintaining the integrity of the research specimens. Specific collection care issues addressed in this volume include the conservation of fluid-preserved collections (Simmons), policies for using and managing ancillary preparations of vertebrate collections (Cato and Schmidly), policy issues relative to destructive testing of archaeological collections (Bohnert and Surovik-Bohnert), and the needs of paleontology collections (Shelton). Even small museums with education-oriented collections have begun to question how to care for their materials (Silvy and Cato).

The purpose and methods of exhibits and educational programming in natural history museums have continued to evolve, reflecting the profession-wide trends. However, the critical state of science education in this country calls attention to the fact that the efforts described in these papers are only the beginning of efforts to improve learning in natural history museums. DeMars provided a general perspective of the state of exhibitions in natural history museums, and Tirrell discussed how traveling exhibits serve a critical goal in the missions of a university-state museum. Efforts by natural history museums to reach out to the communitites they serve are described by Gottfried *et al.* (for improving science education in rural schools), Deisler-Seno and Reader (curriculum-oriented programs), and Patton (loan materials for the classroom).

The contents of this volume should be used to stimulate continued analyses of the philosophies and functions that drive natural history museums. Are the directions and issues discussed herein valid ones? How can they be improved or modified to fit the needs of individual institutions, or the needs of society? The museum profession should never be content to accept the status quo. People change, societies change, and the organizations that serve those societies must change also.

Paisley S. Cato
Clyde Jones

Roles and Functions

Natural History Museums: Directions for Growth
Paisley S. Cato and Clyde Jones, editors
Texas Tech University Press, Lubbock, 1991, iv+252 pp.

THE NATURE OF UNIVERSITY NATURAL HISTORY MUSEUMS

Philip S. Humphrey

Abstract.—A comprehensive natural history museum comprises two museums: an inner scholarly one where the collections are cared for and studied, and an outer museum that makes the knowledge of the inner museum available to the lay public. Research and graduate education, along with care of the collections, are the primary—and for many—the only missions of university natural history museums. Graduate education programs associated with university natural history museums should produce scholar-curators who understand and utilize the outer museum as a means of fulfilling their broader obligations to society.

Events of the past two or three decades have had a profound impact on university natural history museums, broadly defined. There has been a major conceptual and methodological revolution in the field of systematic and evolutionary biology. Nevertheless, biology faculties at many universities consider natural history collections and systematic biology as old-fashioned, backwaters of science in contrast to the modern challenges and promise of chemical molecular biology, genetic engineering, and the like. Even though at long last there is growing public recognition of the enormity of the global biological diversity crisis, in stock-market terms, the job market for systematic biologists continues to reach new lows. As if this were not enough, the larger United States community of museums as represented by the American Association of Museums (AAM) and other organizations does not appear to understand the special nature of university natural history museums and the problems confronting them. Thus, university natural history museums currently face many difficult challenges, some of the most important of which concern their relationships with the larger museum community and with the public in general.

In what follows, I will present a conceptual model for a natural history museum and discuss the special missions of university natural history museums. Following this, I will discuss one of the current problems confronting university natural history museums and present some possible solutions.

A conceptual model for a comprehensive natural history museum.— The unique and distinctive attributes of natural history museums, compared to other kinds of museums, reflect the

5

essential functions of systematic collections in systematic and evolutionary biology. Systematic collections and their associated laboratories and libraries are the basis for cataloging the diversity of living things, understanding their evolutionary relationships, characterizing their past and present distributions and the mechanisms responsible for them, and elucidating the processes that result in organic diversity. Just as libraries and art collections are among the permanent repositories preserving our cultural heritage, so are systematic collections permanent and growing records of our biotic heritage. The preoccupation of natural history museums with organic diversity distinguishes them from other kinds of museums, which—perhaps simplistically—can be said to be concerned with the cultural diversity of human beings, a single species of organism.

A comprehensive, or "full-service," natural history museum actually comprises two museums: an inner scholarly one and an outer public museum. The inner natural history museum consists of the collections and the people who care for and study them. It is a scholarly institute in which, ideally, all the approaches characterizing the history of systematic and evolutionary biology—from alpha taxonomy to biochemical systematics—are practiced to elucidate and characterize biological diversity and the processes responsible for it. The outer museum consists of all those translational devices such as exhibits and public education programs that make the knowledge of the inner natural history museum available to the lay public.

The different publics served by natural history museums.—A comprehensive, full service natural history museum has two publics: a lay public served in specific ways by the outer museum and another public, the international scientific community, served in very different ways by the inner natural history museum.

For a university natural history museum, the "public" of the inner museum is the international scientific community of systematic and evolutionary biologists, as well as students in the graduate programs that educate new scholars in the field. Funding from the Biological Research Resources Program (BRRP) and to a certain extent from the Systematic Biology Program of the National Science Foundation (NSF) is related to the quality of the collections and the abundance and quality of the scholarship and services associated with the collections.

Many university natural history museums funded by the NSF-BRRP are internationally important systematic collections with associated scholar-curators caring for the collections, creating new knowledge, training new systematic and evolutionary biologists, and in various ways serving the national and international scientific communities. These university inner natural history museums, although not serving the lay public through associated outer museums, are nevertheless engaged in realizing the fundamental missions of natural history museums in institutions of higher education.

The outer natural history museum serves the lay public and, in a university, the relevant undergraduate programs of the institution. The AAM definition of a museum (or one similar to it) is used to determine eligibility for funding from the Institute for Museum Services (IMS) and certain other federal sources. The AAM definition in effect represents the lay public in setting forth the requirement that a museum exhibit tangible objects "to the public on some regular schedule."

A comprehensive university natural history museum that can meet the expectations of both its scientific and lay publics through the inner and outer museums obviously is in an advantageous position with respect to eligibility for a broad range of federal support.

The special missions and responsibilities of university natural history museums.—As I discuss the special missions and responsibilities of university natural history museums, please note that we are dealing with matters of emphasis. University natural history museums are by no means unique in their involvement in programs of graduate education. Most, if not all, major U.S. nonuniversity natural history museums have collaborative arrangements with universities with respect to graduate education programs.

Research and graduate education, along with care of collections, are the primary, and for many, the only missions of university natural history museums. Thus, an important element in the strategy of university natural history museums in fulfilling their primary missions is their close relationship with related academic programs of the university. University natural history museum curators often are tenured (or tenure track) faculty members with special responsibilities for undergraduate and graduate teaching and supervision of the graduate programs of students in their charge. In addition, faculty-curators jointly appointed in academic departments participate in shaping the nature, direction, and content of the undergraduate and graduate education programs that ultimately produce scholars qualified for employment in the larger national community of natural history museums.

University natural history museums, through their involvement in graduate education programs, play a crucial role in creating new generations of scholar-curators who are essential for the continued ability of the national community of museums to fulfill its obligations to science and society. In this context, one of the significant failures of graduate programs in systematic biology is that most of the scholar-curators they produce lack an understanding of the broader societal obligations of natural history museums, and have little educational and experiential background in the totality of the complex missions and associated functions of a comprehensive natural history museum.

Why are university inner natural history museums effectively not part of the larger U.S. museum community?—The majority of scholar-curators associated with university natural history museums work in isolation from the larger U.S. museum community, most of them know practically nothing about it, and the little they do know provides little or no encouragement for them to become involved. Why? Virtually all of these people are systematic biologists, using their collections and associated libraries and laboratories to create knowledge about biological diversity. Their public is the international scientific community and those students enrolled in specialized graduate programs. They are engaged in the increase of knowledge about the living world and the diffusion of this knowledge to fellow scientists and students. Although they are deeply involved in education, they are using the educational principles and practices characteristic of universities, not of the outer museum, and their principal educational objective is to produce new scholars in their particular fields of specialization—not to educate the general public.

Most scholar-curators in university natural history museums are inner museum oriented, whereas the larger U.S. museum community, its professional organizations, and certain of its sources of federal funding are largely outer museum oriented.

There is very little basis for communication between the two. Very few scholar-curators in university natural history museums belong to the AAM and most are at best only dimly aware of its existence. It is no wonder that the AAM does not fully recognize and understand the special problems of university natural history museums, or of inner natural history museums in general. These museums, in large part by their own choice, are not adequately represented in the deliberations of the AAM.

At a time when there is growing national and international concern about the global biodiversity crisis, there are very few universities with the collections and associated faculty required to inspire undergraduate and graduate students to take up careers in systematic and evolutionary biology. If during the next decade plans are developed to inventory the rapidly disappearing tropical forests and other environments of the world, there surely will be a serious shortage of trained systematic biologists to implement these plans. The principal national resource for training new systematic biologists is a small group of universities with natural history collections and research and graduate education programs in systematic and evolutionary biology.

As a corollary to this, the public needs to be much better informed about the nature of the biosphere, our civilization's life support system, and the frightening rate at which it is being modified and destroyed. Natural history museums are one important way of accomplishing this, and scholar-curators, most of them trained in close association with university natural history museums, should have greater involvement in the outer natural history museum as a means of realizing their critical obligation to the lay public and to decision-makers at all levels of government and private enterprise.

Some thoughts about what might be done about the problem.—As I have stated, most scholar-curators in university natural history museums are isolated from the national and international outer-oriented museum communities and thus are neither contributing to them nor benefiting from them. This is a two-way street on which there is little traffic at the moment. Most faculty-curators associated with university natural history museums probably do not even realize this is a problem. The larger community of mostly outer-oriented museum people, when they think about this situation, do not understand why inner natural history museum people are not more involved with the increasingly active, lay public-oriented American museum movement.

Before discussing possible solutions, we must understand why the problem exists in the first place. The principal reason is the nature of the reward system for systematic biologists.

What are the criteria for employment as an inner-museum curator in a natural history museum? You must be a productive (or promising) scholar with a career research interest in the group of organisms represented by the systematic collection with which you wish to be associated and you must have a Ph.D. Your graduate education taught you to be a scholar and you learned that getting on the fast track for jobs in natural history museums and universities requires scholarly productivity and involvement in national and international professional societies. Moreover, once employed as a scholar-curator, promotion, tenure, salary increases, and success in obtaining federal grants also depend on scholarly productivity and professional recognition by the national community of systematic and evolutionary biologists.

Promotion, tenure, salary increases, research grants, and professional recognition by the scientific community for a university natural history museum curator are not obtained by involvement in public service activities or in museum-related organizations such as the AAM. The professional rewards for scholar-curators in most university natural history museums result from inner, not outer, museum activities and from involvement with the scientific community, not with the mostly public service-oriented community of museum people.

What are some of the possible solutions to this problem? Before describing them, let me say at the outset that I do not advocate demolishing the existing reward system for scholar-curators employed in university natural history museums; it encourages scholarly excellence and productive graduate education programs. The solutions are simple in principle, but in practice possibly difficult to implement, and especially so in museums plagued by limited resources, which is the case for the majority of university natural history museums. The following solutions are specifically addressed to university natural history museums, recognizing that some of them already may be implementing some or all of the solutions I will discuss.

1. The reward system for faculty-curators should be expanded to include additional rewards for involvement in museum public service activities.
2. Those university natural history museums whose missions are now limited to inner museum programs might consider expanding their missions to include certain public service activities, even on a very limited basis, and to involve faculty-curators and qualified graduate students in these activities. Many outer museum activities (for example, school services, workshops) can be designed and managed so they generate fee income to defray some of the costs, but some commitment of permanent staff is required.
3. Doctoral level graduate programs in systematic and evolutionary biology should address the roles of systematic biologists in improving public understanding of science and should provide opportunities for graduate students to become actively involved in school service and other public education activities.
4. The AAM, Museum Services Board, the NSF, and the Association for Systematics Collections might profitably address development of programs and funding mechanisms that would encourage scholar-curators in university natural history museums, along with their associated graduate programs in systematic biology, to become more actively involved in educational and other public service activities designed to improve public understanding and support of systematic biology and natural history museums.

One clear objective should be to develop individual and institutional reward systems that would encourage increased involvement by faculty-curators of university natural history museums in outer museum activities and in the AAM and related organizations to the ultimate benefit of all concerned.

Natural History Museums: Directions for Growth
Paisley S. Cato and Clyde Jones, editors
Texas Tech University Press, Lubbock, 1991, iv+252 pp.

WHAT IS A STATE MUSEUM OF NATURAL HISTORY?

Joshua Laerm and Amy Lyn Edwards

Abstract.—There are 23 officially recognized state museums of natural history or state museums whose missions relate in large part to natural history. A survey of these museums was undertaken to define their specific missions or roles, ascertain the relative importance of these missions, compare the administrative base of these museums and obtain information regarding funding sources.

Explicit in the stated missions of these museums are four areas of responsibility: collections, their acquistion, conservation, and management; research; education; and service. The survey indicates that collections and collection-related research are their primary focus, with educational and service activities a natural extension.

The majority (13) of these museums are administered through various state agencies. Nine are administered within the respective states' large research universities. One is private.

State museums are funded largely through state appropriation. Four are supported entirely through state appropriations. Seventeen are supported by a combination of state appropriation and independent sources (contracts, grants, gifts). One is private, though supported to some extent by state appropriation.

Ours is a nation of museums. There are over 6000 of them in all. In certain respects they have been inspired by what might almost be considered a Jeffersonian egalitarianism. For, since the late 18th and early 19th centuries, our nation's museums have been evolving as intellectual and social mechanisms for the democratization of knowledge and culture.

Individual museums have oriented and grown for various reasons. They are guided by special interests and special needs. Each has its own constituency. There are national museums, state museums, city museums, and local museums. There are museums of science and technology, museums of art, history, and natural history. There are zoos, nature centers, and botanical gardens. Some of them are large, some of them are small. Each is important in its own way. Together, these museums represent our cumulative efforts to preserve our past, our natural heritage, and the most beautiful and creative examples of the human experience.

One of the most prominent groups of museums in our nation are natural history museums. They include many of our largest, oldest, most popular, and visible museums. As a group, natural history museums are as varied as are other types of museums. They vary in terms of their governance. There are, for example, natural history museums associated with public and private universities and colleges (some are even administered within academic departments); some are public museums established at the national, state, and local levels, and some are private museums of natural history. The missions and roles of these museums are varied as well. Some of them have strictly research, educational, or service functions. Many, if not most, have a combination of roles. These museums vary as well in terms of sources of funding and support.

This paper reports on the results of a survey and review of the missions, administration, and funding of one particular groups of natural history museums–state museums of natural history.

METHODS

To define the conceptual and functional roles provided by officially recognized state museums of natural history, a survey of these museums was undertaken. Twenty-three state museums were identified from information provided by the American Association of Museums, and direct contact, as being "officially" recognized state museums of natural history or state museums that relate largely, or in significant part, to natural history. A listing of these museums is provided in Tables 1, 2, and 3. State museums that relate primarily or exclusively to cultural or social history were excluded from this survey.

Four specific questions were submitted to each museum through their director or other appropriate administrative authority.

1. What are the stated missions or responsibilities of your museum?
2. What is the relative importance of the various roles that your institution provides?
3. What is the administrative relationship of your institution to other state agencies?
4. How is your institution funded?

The specific responses to the survey are provided in the Appendix. What follows is a summary of that survey.

RESULTS AND DISCUSSION

Missions of a State Museum of Natural History

Explicit in the stated missions of all state museums surveyed are essentially four primary areas of responsibility: (1)

TABLE 1.—*A listing of state museums of natural history surveyed.*

1. University of Alabama State Museum of Natural History
2. Alaska State Museum
3. The Connecticut State Museum of Natural History
4. Florida State Museum
5. Bishop Museum (Hawaii)
6. Idaho Museum of Natural History
7. Illinois State Museum
8. State Historical Museum of Iowa
9. Maine State Museum
10. Mississippi Museum of Natural Science
11. University of Nebraska State Museum
12. Nevada State Museum
13. New Jersey State Museum
14. New Mexico Museum of Natural History
15. New York State Museum
16. North Carolina State Museum
17. Oklahoma Museum of Natural History
18. The State Museum of Pennsylvania
19. South Carolina State Museum
20. Texas Memorial Museum
21. Utah Museum of Natural History
22. Virginia Museum of Natural History
23. Thomas Burke Memorial Washington State Museum

collections, their acquisition, conservation, and management; (2) research; (3) education; and (4) service.

Collections

Maintaining collections.—This is the most basic aspect of all state museums of natural history. Essentially all the museums surveyed are major repositories, if not the definitive repositories, of collections of objects, artifacts, and specimens that relate to the natural history of their respective states. Several of these museums maintain collections of regional, national, or world significance as well. It is clear that these museums view their collections as a resource, data base, and focus upon which other functions depend.

1. The collections are viewed as important for the basic and applied research programs in which most of these museums are involved.

2. They represent the raw material from which exhibit programs are designed and educational programs developed.
3. The collections, their associated data bases, and the expertise of their staffs provide the basis for the specialized technical and informational services for which these museums are broadly recognized.

Furthermore, discussions with several of the administrators of these museums indicate that collections themselves were the basis for the formation, recognition, or both, of these museums as "state" facilities. Several museums were formed with the responsibility of developing collections that relate to the natural history of their state. Alternatively, other state museums were formed through the recognition of a pre-existing museum with important collections relating to that state.

Acquisition.—Most of the museums surveyed specifically mention active acquisition and collecting as a primary responsibility. Growth of collections results from a museum's inherent role as a repository of important objects, artifacts, and specimens. Growth results from a museum's attempt to serve the needs and interests of the present generation while anticipating that of future generations. On a day-to-day basis, however, museum collections most often are driven by any number of very specific research, educational, and service demands placed upon them, as well as the specialized interests of their research and curatorial staffs.

Conservation.—Conservation means care and maintenance. Its goal is preservation. Merely entrusting an object to a museum does not mean that it will be preserved in perpetuity. The conservation of large and diverse natural history collections requires sophisticated technological solutions to a wide variety of destructive processes and agents that threaten artifacts and specimens. Even rocks and bones undergo serious deterioration if not properly protected. Metallic archaeological artifacts are subject to atmospheric oxidation. Light, temperature, and humidity can have devastating effects on many types of biological specimens. An infestation of insects essentially can destroy a collection of millions of biological specimens in a matter of weeks. Storage practices that were common just a few decades ago are now outdated or even considered deleterious to collections. The types of storage cases, the ink, the very paper upon which records are kept is of serious concern in a museum.

Management.—Today, collection management means not merely cataloging and arranging artifacts and specimens but also managing information. Museums of natural history are as much repositories of information as they are repositories of objects. And, as such, they must be able to respond to requests for that information. This creates a responsibility for effective data management and retrieval. In many cases informational services are becoming such an important aspect of many museums that growth of their collections is driven by the service demands placed upon them.

In today's museums, the mere existence of a collection and catalog of specimens is no longer satisfactory to meet the needs of scientists, educators, local, state, and federal agencies, and the general public. To a great extent, the value of a museum's collections now depends upon the availability and accessibility of information associated with the objects that make up the collections.

State museums have a new responsibility for effective data management and information retrieval. The synthesis of voluminous information, the cross referencing of data and specimens, maintaining and updating inventory site files, historical and current location and distributional records, intercollection correlations for collections numbering in the hundreds of thousands, even millions of specimens, can be a data manager's nightmare. State museums are becoming increasingly dependent upon computer technology. Information requests that previously required the full-time labor of days or weeks to retrieve from catalogs and files may now be available almost instantaneously by computer. However, the simple act of putting the information of a single collection into a computer can require years of labor.

Research

The major museums of natural history in our nation are important research institutions. State museums of natural history are not exceptions. Essentially all the museum administrators surveyed indicated that research related to the culture, history, and natural heritage of their respective states is a primary function of their institutions. The basic function of research in any museum is to increase knowledge. This is part of a museum's responsibility to its constituency. It is through research and the knowledge gained from it that museums are able to provide many of the highly specialized technical, informational, and educational service functions for which they are broadly recognized and valued.

Many of the state museums surveyed represent an important research resource for other public administrative units within their respective states. These museums provide floral, faunal, archaeological, and geological surveys; environmental impact assessments; and technical, informational, and consulting services in conservation and natural resource management. The individual collections of these museums and the data bases of information and knowledge gained from research represent a significant resource for their states. Cummulatively, these museums provide an irreplaceable source of knowledge of national and international significance.

In many cases, museums of natural history are the only institutions in which certain types of research are conducted. A number of the museums surveyed have a high degree of national and international visibility and recognition based upon the quality and scope of the research programs of their staff. This is particularly true in the areas of systematics and biodiversity. Of the estimated five to 30 million species of organisms in existence today less than two million have been described. The earth is losing species at an average rate of 100 per day. In the next 50 years, 30 to 50 percent of all extant species will become extinct. Present extinction rates of the earth's biota can be estimated at a thousand times that of any time since the Cretaceous. The issues of systematics collections, biodiversity, and the maintenance of information contained within the genetic code of these organisms are survival issues for mankind. In every way and with the complete agreement of the national and international community, these state museums are in a unique, if not critical, position to address these problems on a local and regional basis.

Education

The importance of museums as educational institutions has long been recognized in the United States. The survey respondant from these state museums of natural history clearly indicate a strong commitment to education. In many respects, these museums and others are in the forefront of our national educational system. Museums of natural history represent unique institutions for the transmission of knowledge, understanding, appreciation, and the development of aesthetic responsibilities regarding the natural world and man's place in it. To quote the respondents, museums "illustrate, exhibit, and explain." They "enhance understanding," "improve opportunities for aesthetic and learning experiences," and "provide for cultural enhancement." Museums do good things!

Review of information regarding the respective museums' educational programs indicates they range from little more than static exhibits to broadly integrated and sophisticated in-house and extension educational services to a wide variety of segments of their constituency. The fact that the importance of education in museums of natural history (and other museums as well) has increased in recent years was expressed several times in the survey responses and in follow-up discussions with several administrators. The reason for this increasing emphasis is that the public, a large part of their constituency, views their state museums as educational institutions. As a result, the visibility and success of a museum's educational programs are in large part determining factors in the ability of museum adminstrators to fund other functions and programs. State museums not only have the capacity for unique educational services, they have the responsibility to provide them.

Service

Most of the museums surveyed indicate "service" as one of their stated missions. Unfortunately, none of them define what is meant by the term. In some respects, it is clear that virtually everything a museum does relates to service. Therefore, it is difficult to separate the concept of service from a museum's other roles. State museums are basically service institutions.

One possible reason why state museums of natural history recognize service as such an important component of their total mission is that much of what these museums do has broad practical applicability. Museums of natural history typically have a comparatively broader disciplinary base encompassing a broader range of fields than do other types of museums. The information and knowledge related to natural history, in many cases, is of considerable practical value to numerous local, state, and federal agencies, as well as the general public. For example, the types of services provided by the various state museums surveyed include: archaeological, biological, and geological surveys; environmental impact assessments; scientific technical assistance; informational services; publications; collection loans; specialized reference libraries; specialized classes and training programs; exhibit design, development, and loan services; film and audiovisual libraries; and various educational programs. This list goes on. But these are basically services that relate ultimately to the roles that these museums fulfill as repositories of objects and information, research institutions, and educational facilities. Service, it seems, is simply inherent in what they do.

Relative Importance of Roles

This survey clearly indicates that collections and collection-related research are the primary responsibilities of state museums of natural history. Few of the respondents ranked education and service equal in rank to their role as a repository and research institution.

This survey supports the widely held view in the museum community that state museums of natural history are important repositorial and research centers. It is obvious that these museums of natural history function to collect and preserve the tangible record of their respective natural heritages. It is through their research programs, however, that they can make significant contributions to the knowledge of their states, and through their cumulative efforts, ultimately to our nation.

Education and hence understanding and appreciation must ultimately be based upon knowledge. These state collections, their associated data bases, and the results of their research efforts are the basis for that knowledge. Thus, it is perhaps appropriate that education and other services are a natural extension of these basic functions. It would be difficult to provide significant and meaningful educational experiences, much less specialized services, regarding the natural history of a state that had no body of knowledge or a staff of knowledgeable experts to draw upon.

Similarly, all the museums surveyed indicate a very strong commitment to education and other specialized services. In so doing, they recognize that it is the educational and service roles of their institution that gives it meaning, direction, and purpose.

Two quotes from *Museums for a New Century, A Report on the Commission on Museums for a New Century of the American Association of Museums* (1984) perhaps put it in better perspective.

The act of collecting and preserving objects is at the center of the museum domain. Just as important is the use of collections to advance knowledge and understanding and thus it is through research, education and exhibition that museums make their collections available.

The mission of museums as institutions grows out of tradition and is shaped by public need. It is to collect and preserve the evidence of the natural and physical world and of human accomplishment, and to use those collections and related ideas to contribute to human knowledge and understanding.

Museum professionals have long recognized that the public's view, appreciation, and even support of museums is based primarily upon their "up front" activities—their education and service activities. In general, the public does not understand the underlying importance of the collections or research programs. Often head are statements such as: "What good is it if it is stuck in storage cases where no one can see it?" "Why do you need 3000 examples of the same thing?"

It is perhaps to the credit of these state museums and their legislative constituencies that they recognize the underlying concept of what a museum is and what it does. Museums are important because they are repositories of knowledge and information. They are fun, entertaining, and educational. But museums are serious places also; they are centers of knowledge and information that are important in our understanding and appreciation of the natural world around us. Ultimately, it is the collections and the research conducted upon them that permit us to understand the "nature of things."

Organizational Structure

Table 2 outlines the administrative base or supervisory agency for the respective state museums of natural history surveyed. The majority (12) of these museums are administered though various state agencies or departments. It is interesting to note the variation in relationships that museums of natural history bear to supervisory agencies within their respective states: agriculture, archives, conservation, cultural services, education, energy, history, natural resources, state, wildlife. It is inviting to suggest that it is a reflection on the varied roles that such museums bear to society that museums of natural history can serve the interests or needs of so many different state agencies. In this view it would appear to be an overwhelming endorsement of their value.

A large proportion of these museums (9) are associated
with their state's major educational institution and, in fact,
are administrative units within them. Several of these would
probably be ranked within the top 25 natural history museums
in the nation regarding the significance of their collections
research and service programs. This is due in large part to
their relationships to major research oriented universities.

The single exception to state or university administered
state museums is the Bishop Museum of Hawaii. This is a
private museum only recently recognized as the State
Museum of Hawaii.

Funding

Table 3 outlines the funding source(s) for the museums
surveyed. State museums of natural history are supported
largely by state appropriations through the appropriate
governing authority. Five of the museums are supported totally through state appropriation. The remaining museums
are supported by a combination of state appropriations and

TABLE 2.—*Administrative base or supervisory agency for state museums of natural history.*

Museum	Supervisory Agency
Alabama	University of Alabama, Academic Affairs
Alaska	State Board of Education
Connecticut	University of Connecticut, President
Florida	University of Florida, Academic Affairs
Hawaii	Bishop Museum, Private
Idaho	University of Idaho, Academic Affairs
Illinois	State Department of Energy and Natural Resources
Iowa	State Historical Department
Maine	State Department of Education and Cultural Services
Mississippi	State Department of Wildlife and Conservation
Nebraska	University of Nebraska, Research and Graduate Studies
Nevada	State Department of Museums and History
New Jersey	Department of State
New Mexico	State Division of the Office of Cultural Affairs
New York	State Department of Education
North Carolina	State Department of Agriculture
Oklahoma	University of Oklahoma, Research
Pennsylvania	State Historical and Museum Commission
South Carolina	State Museum Commission
Texas	University of Texas at Austin, Provost
Utah	University of Utah, Academic Affairs
Virginia	Secretary of Natural Resources
Washington	University of Washington, Arts and Sciences

TABLE 3.—*Funding sources for state museums of natural history.*

Museum	Budget Year	Funding Source(s)
Alabama	1987	State appropriation (35%) contracts and grants (45%) gifts (20%)
Alaska	1987/8	State appropriation (88%) gifts (12%)
Connecticut	1986/7	State appropriation (60%) gifts (30%) grants (10%)
Florida	1985/6	State appropriation (72%) other 38% *
Hawaii	1987	State appropriation (29%) contracts, grants, gifts (71%)
Idaho	1985/6	State appropriation (70%) contract, grants, gifts (30%)
Illinois	1986/7	State appropriation (70%) contracts, grants, gifts (30%)
Iowa	1986/7	State appropriation (80%) other (20%) *
Maine	1986/7	State appropriation (90%) independent sources (10%)
Mississippi	1986/7	State appropriation (100%)
Nebraska	1986/7	State appropriation (80%) independent sources (20%)
Nevada	1986/7	State appropriation (80%) independent sources (20%)
New Jersey	1986/7	State appropriations (80%) Independent sources (20%)
New Mexico	1986/7	State appropriation (60%) independent sources (40%)
New York	1986/7	State appropriation (100%)
North Carolina	1986/7	State appropriation (100%)
Oklahoma	1987/8	University appropriation (75-80%) state, grants, gifts (20-25%)
Pennsylvania	1987/8	State appropriation (74%) gifts (21%) grants (5%)
South Carolina	1987/8	State appropriation (80%) private (20%)
Texas	1987/8	State appropriation (70%) gifts and grants (30%)
Utah	1986/7	State appropriation (70%) independent sources (30%)
Virginia	1988/9	State appropriations (98%) grants, gifts, membership (2%)
Washington	1986/7	State appropriation (2%) independent sources (50%)

*gifts; donations, admission fee, gift store revenue

independent sources (contracts, grants, gifts, endowments, loans, and so forth). In only four cases (Alabama, Connecticut, Hawaii, and Washington), was state support less than 50 percent. The level of state support in the remaining museums ranged from 70 percent to 93 percent.

Information regarding the actual level of funding was not solicited as it is not instructive in the absence of information regarding the relative size of physical facilities, staff, and programming.

APPENDIX.—*Survey responses to questionnaire.*

University of Alabama State Museum of Natural History

I. Stated missions:

"The University of Alabama State Museum of Natural History is dedicated to the collection and study of materials and artifacts pertaining to the natural environment of Alabama and the southeastern United States; the Museum supports the development and dissemination of knowledge derived there through programs in research, exhibition, education and publication.

"The museum supports the University's mission of teaching, research and service through public programs for children, teachers and adults, the maintenance of collections to support student and faculty research in the natural sciences, and through research conducted by its own staff. The University's commission to enhance education is fulfilled by the Museum's support of science education for teachers, exhibition programs in geology, archaeology, biology and ethnology on campus, at Mound State Monument, and at exhibit sites throughout Alabama. The Museum also serves the public good through its program in salvage archaeology in Alabama and the southeastern United States."

(Source: letter from D. Jones, September 10, 1987).

II. Relative importance of roles:

"The Museum's role includes repository responsibility for natural history and archaeological materials, research, service and public education.

Current focus is on collection assessment and curation, archaeological research, and educational programs within the museum and as outreach programs."

(Source: Same as above)

III. Organizational structure:

"Approval was given to the establishment of University of 'Alabama Museums' to serve as the principal organization for the State Museum of Natural History and the Paul W. Bryant Museum, with their subordinate units and others thay may be added in the future."

(Source: Same as above)

IV. Funding:

"Funding for the Museum is provided by a legislative line item, separate from the University's."

(Source: Same as above).

Alaska State Museum

I. Stated mission:

"The Division of Alaska State Museum is a state governmental agency, charged by statute with the identification, collection, preservation, and public exhibit of

Alaska's material history, cultures, and natural history; the Division interprets and disseminates knowledge of the history of the state, its people, and its resources and supports other statewide efforts in this direction. The Division is also charged with the responsibility to support the growth, development, and excellence of other museums within Alaska."

(Source: Draft, long range plan, T. Lonner, December 24, 1987)

II. Relative importance of roles:

"The sustaining core of the State Museums is the collection. The most sizable proportion of the Museums' fiscal and human resources is devoted to understanding, preserving, securing, conserving, enhancing, enlarging, and displaying the collection to the public.

The Division of Alaska State Museum carries a central responsibility to communicate the history and culture of Alaska to present and future generations of residents and visitors through the display of material objects collected by or loaned to Alaska. This responsibility guides the organization of work; the state museums develop their themes, research projects, exhibit projects, and focused collecting."

(Source: Same as above)

III. Organization structure:

"The State Board of Education and the Commission of Education Review all regulations, administrative procedures and divisional policies. The Division of Alaska State Museum maintains two separate museum facilities and provides support for various local museums and schools across the state."

(Source: Same as above)

IV. Funding:

"The Division of State Museums, as a governmental agency, is dependent principally on state general funds."

(Source: Same as above)

The Connecticut State Museum of Natural History

I. Stated mission:

"The Museum of Natural History at the University of Connecticut shall be the state museum of natural history. The museum shall (1) acquire, collect, preserve, research and interpret living, extinct and fossil organisms, anthropological and archaeological specimens, rocks and minerals, with preference to those native to this state; (2) preserve related scientific equipment, instruments and data; and (3) prepare public exhibits at the museum and educational exhibits and programs that may be used by colleges, universities, schools, libraries, institutions, appropriate state agencies or other public organizations. There shall be an office of archaeology at the museum. The museum shall be an independent entity within the University of Connecticut for administrative purposes only."

(Source: Senate bill No. 341 public Act No 85-563, July 1, 1985
and Substitute House Bill No. 5166 public Act No. 87-466, 1987)

II. Relative importance of roles:

"Currently outreach education programs that will help promote better science education, are deemed more important than research or collections."

(Source: Letter from C. Rettenmyer, September 7, 1987)

III. Organizational structure:

"The board of directors of the museum of natural history shall consist of the following: The commissioners of education, environmental protection and agriculture or their designees, the director of the State historical commission or his designee and not more than eleven members appointed by the president of the

University of Connecticut, of which seven shall be professors, at least one from each of the following fields: Anthropology or archaeology, geology, vertebrate biology, invertebrate biology, botany, systematic biology and any other field the president deems appropriate. The board of directors shall be responsible for the planning and establishment of the state museum of natural history and shall recommend a director who shall be appointed by the president of the University of Connecticut."

> *(Source: Senate Bill No. 341, Public Act No. 85-563, July 1, 1985).*

IV. Funding:
"Funding is provided by the legislature as a line item, and is not part of the University's budget."

> *(Source: Letter from C. Rettenmyer, September 7, 1987)*

Florida Museum of Natural History

I. Stated missions:
"The functions of the Florida State Museum [now called the Florida Museum of Natural History] located at the University of Florida, are to make scientific investigations toward the sustained development of natural resources and a greater appreciation of human cultural heritage, including, but not limited to, biological surveys, ecological studies, environmental impact assessments, in depth archaeological research, and ethnographic specimens and materials in sufficient numbers and quantities to provide within the state and region a base for research on the variety, evolution, and conservation of wild species; the composition, distribution, importance, and functioning of natural ecosystems; and the distribution of prehistoric and historic archaeological sites and an understanding of the aboriginal and early European cultures that occupied them. In addition, the museum shall develop exhibitions and conduct programs which illustrate, interpret, and explain the natural history of the state and region and shall maintain a library of publications pertaining to the work as here-in provided. The exhibitions, collections, and library of the museum shall be open, free to the public, under suitable rules to be promulgated by the director of the museum and approved by the University of Florida."

> *(Source: Florida State Museum Statutory Authority (1983) 240.515)*

II. Relative importance of roles:
"The FSM is basically a research oriented museum. Collections oriented research receives the bulk of the support budget. We have both the statutory and moral obligation to serve as a regional and national repository for collections in our realm of expertise, i.e. Florida and circum-Caribbean natural history. Many of our collections have been designated national and regional resources which further intensify our repositorial obligations. We rank service and education collectively behind research and collections management. However, we recognize the urgency of expanding our contributions to statewide services. We are placing increased emphasis on providing statewide services in the areas of research , public service and education, using both our collections and our staff expertise."

> *(Source: Letter from Dr. C. Shaak, 28 September 1987)*

III. Organizational structure:
"The FSM ranks as a college within the University of Florida Education and General budgetary framework. The director of the museum reports directly to the Provost and serves on the Council of Academic Deans. The museum is divided into four operational units: the Director's Office and the departments of Natural Sciences, Anthropology, and Interpretation. The museum faculty are regular

university faculty with tenure earned in the museum. Each of the departments has a faculty chairman."

(Source: Same as above)

IV. Funding:
"The budget is included as a line item entry in the University of Florida Education and General budget."

(Source: FSM News. 1985. VOL XIV, No. 1)

Bishop Museum

Information regarding enabling legislation and details of organization not available.

I. Stated missions:
"The mission statement of Bishop Museum: Bernice Pauahi Bishop Museum is dedicated to gathering, preserving, studying, and sharing knowledge of the cultural and natural history of Hawai'i and the Pacific. The museum seeks to document the past, understand the present, and prepare for the future through research, conservation, publication, education, exhibition, and other programs designed for enjoyment and enlightenment."

(Source: Letter from M. Altiery, September 27, 1988)

II. Relative importance of roles:
"The museum seeks to achieve a balanced approach between the three line functions of Research and Scholarly Studies, Collections Management, and Public Programs. While Research and Scholarly Studies receives more funding than the other two areas, proper care of our collections is our most sacred public trust. In recent years, more resources have been invested in Public Programs to reach a broader audience and expand our constituency."

(Source: Same as above.)

III. Organizational structure:
"Bishop Museum is a private, non-profit institution that has been designated by the state legislature as the official State Museum of Natural and Cultural History. The designation does not give the State any management rights in the museum which remains autonomous.

The museum is governed by a board of directors and managed by the chief executive officer who is the president of the corporation and director of the museum."

(Source: Same as above.)

IV. Funding:
"Bishop Museum is funded by monies from our endowment fund, and from gifts, grants, contributions, admission fees, sales in our gift shop, and contract services provided by our research department. We were designated by an act of the legislature as the State Museum of Natural and Cultural History just this year, so this is the first year in which we will be receiving any substantial assistance from the state government. We expect this assistance to continue indefinitely, but not necessarily at the same level."

(Source: Same as above.)

Idaho Museum of Natural History

I. Stated missions:
"The Idaho Museum of Natural History exists to preserve, to increase and disseminate knowledge of natural history of the State of Idaho and the northern Inter–mountain West and by doing so enhance in the citizens of Idaho an understanding of and delight in their natural heritage. The Museum's specific areas of interest encompass the anthropology, botany, geology, paleontology, and zoology

of Idaho and the northern Inter–mountain West. The communities served by the Museum include the Idaho State University Community, the citizens of the State of Idaho, and the national and international community of students and scholars."

(Source: Role and Mission Statement, Idaho Museum of Natural History, 1986)

II. Relative importance of missions:
"The Museum fulfills the various roles about equally between repositorial, research and education. The service role fits in with research and education. We are currently expanding the role of education at the museum, but not at the expense of other roles."

(Source: Letter from W. Steffan, 4 September 1985; 28 September 1987)

III. Organizational structure:
"The Museum was designated the Idaho Museum of Natural History by governor's proclamation in 1977. Organizationally, the Idaho State Museum is a unit of Idaho State University. It is equivalent to a department with the director answering to the Vice President for academic affairs."

(Source: Letter from W. Steffan, 4 September 1985)

IV. Funding:
"Two thirds of their funding is provided by state appropriation. The other third comes from contracts, grants and gifts."

(Source: Same as above)

Illinois State Museum

I. Stated missions:
"(1) Selectively expand and improve the quality of collections significant to Illinois, resources that provide solid bases for research, exhibition, and education.

(2) Ensure that collections receive proper care through sound curatorial and conservation practices that meet recognized professional standards.

(3) Conduct basic research in areas appropriate to the collections in order to optimize the unique information potential of collections, thus contributing directly to the Museum's educational mission.

(4) Exhibit collections in a manner that enhances and improves the public opportunity for aesthetic and learning experiences and, ultimately, improves the quality of life.

(5) Disseminate information on the State's natural history, anthropology, and art to a broad public spectrum through a variety of educational programs and activities.

(6) Plan strategically in order to serve the immediate needs of the present generation while insuring preservation of the Museum's resources."

(Source: Illinois State Museum Statement of Purpose, 13 November 1985)

II. Relative importance of roles:
"In approximate relative order as indicated above."

(Source: Same as above)

III. Organizational structure:
"The Museum is a division of the State Department of Energy and Natural Resources. The Museum Director reports to a cabinet officer of the governor, who in turn serves as an ex-officio member of the Museum Board. The Museum's Board is composed of 11 individuals representing the natural sciences, anthropology, art, and business. Board Members are appointed by the Governor and confirmed by the Senate. Administratively, the Museum is organized under the Director and three Assistant Directors who serve as Section Heads."

(Source: Same as above)

IV. Funding:

"Funding is through state appropriation to and through the Department of Energy and Natural Resources and through grants and contracts."

(Source: Pers. comm. by B. Styles, 25 May 1988)

State Historical Museum of Iowa

I. Stated mission:

"The State Historical Musem of Iowa is obligated to assist its citizens in identifying themselves, individually and collectively, in space and time. The Iowa State Historical Department is the State agency created and empowered by the General Assembly as the official trustee of Iowa's human heritage, with the responsibility and authority to ...

— identify, record, collect, and preserve the manifestations of Iowa's history, including, but not limited to, documents, printed materials, images, artifacts, and historic sites;

— interpret and disseminate Iowa history through exhibits, educational programming, printed materials, and other media;

— conduct, stimulate, produce, and provide the resources and opportunities for scholarly research and publication of Iowa history;

— promote and coordinate the teaching, appreciation, and understanding of Iowa history with and through educational institutions and other individuals, organizations, and agencies;

— serve as the primary reference and resource agency on Iowa history for all governmental agencies;

— identify, collect, preserve, and make accessible the official records and documents of state, county, and municipal government;

— advocate the preservation, stewardship and use of Iowa's historical resources; and

— obtain and efficiently manage the resources necessary to effectively fulfill the Department's mission."

(Source: Trustees for Tomorrow, Iowa St. Hist. Dept., January 1986)

II. Relative importance of roles:

"The museum has ranked collection, research and education equally in importance. Within each role they have evaluated and ranked objectives. The preservation and inventory of the collections, promotion of using the collections for research and publication, and promoting public education are all essential parts of their ten year plan."

(Source: Same as above)

III. Organization structure:

"The Governor appoints a Director of the Dept. of Cultural Affairs, who serves at his pleasure. The Director of the Dept. of Cultural Affairs appoints an Administrator of the State Historical Society of Iowa, who serves at the director's pleasure. The Administrator then appoints four Bureau Chiefs. One of the four Bureau Chiefs is the head of the State Historical Museum."

(Source: Same as above)

IV. Funding:

"Approximately 80 percent of the Museum's budget comes through appropriations from the State of Iowa."

(Source: Same as above)

Maine State Museum

I. Stated missions:

"To institute and maintain a program of systematic collection in the several fields of museum activity with particular emphasis on those fields relating to the State of Maine;

To preserve, identify, catalog, document, and safeguard the collections of the museum;

To carry on research to increase knowledge in several fields of museum activity and to provide interpretive and informational services, including the dissemination and recording of information gathered through research;

To provide reference services to individuals and to local history, natural history, scientific and other groups and societies interested in museum activities;

To exhibit the collections of the museum including permanent and temporary exhibits and to provide a loan service for films, artifacts, specimens and other exhibits of the museum to such persons, groups and schools and under such terms and conditions as determined by the director.

To provide ancillary museum services such as operation of a museum library, consultation concerning museum activities, sale of publications, provision of speakers, participation in special events, and other activities which will promote the effectiveness of the museum."

(Source: Enabling legislations, State Museum, 1965)

II. Relative importance of roles:

"The mission of the Maine State Museum, . . . as set forth in our enabling legislation, clearly includes a mandate to preserve and exhibit the environmental . . . richness of the state; a mandate which might be interpreted as requiring our development of major research reference exhibition collections."

(Source: Letter from R. Kley, 3 September 1985)

No particular priorities were expressed.

III. Organizational structure:

"The museum functions as an operating bureau of the State Department of Educational and Cultural Services. The director is appointed by the Maine State Museum Commission with the approval of the Commissioner of Educational and Cultural Services. The Maine State Museum Commission consists of 15 members, especially qualified and interested in the several fields of museum activity, appointed by the Governor. The Commission meets regularly to formulate policies and to exercise general supervision of Museum activities. Standing committees work with the director in the continuing development of long range plans."

(Source: State Museum Bureau Policy Statement, 1972)

IV. Funding:

"The institution is funded, almost without exception, through state appropriations."

(Source: Same as above)

Mississippi Museum of Natural Science

I. Stated missions

"The purpose of the Mississippi Museum of Natural Science is to increase knowledge of the natural sciences, particularly as they pertain to the conservation of Mississippi's wildlife resources, and to disseminate this knowledge among mankind.

    A. The Museum will increase knowledge by:

        1. Collecting specimens and data of Mississippi wildlife resources—plant life and animal life—past and present, and related subjects.

2. Providing proper storage and care for the collected specimens and data, in keeping with professionally accepted Museum standards and practices.

3. Conducting research on these collections and insuring, in accordance with approved procedure, the availability of the collections to the public for research, educational and exhibition purposes.

B. The Museum will disseminate knowledge through:

1. Interpretive exhibits, formal and informal teaching, symposiums, workshops, meetings, programs, public addresses, and the news media.

2. Scientific, educational, and special publications.

3. Extension Services and other special projects that promote the conservation of Mississippi's wildlife resources and advance Mississippians knowledge of the natural sciences."

(Source: Statement of Governing Authority,
Mississippi Commission on Wildlife Conservation, October 21, 1980)

II. Relative importance of roles:

"In relative sequence of order as presented in Part I, above."

(Source: Pers. comm. B. E. Gandy, September 4 1985)

III. Organizational structure:

"The Mississippi Museum of Natural Science is a Division of the Bureau of Fisheries and Wildlife, Department of Wildlife and Conservation."

(Source: Statement of Governing Authority,
Mississippi Commission on Wildlife Conservation, October 21, 1980).

IV. Funding:

"The Museum is funded 100 percent by the Mississippi Legislature through the Department of Wildlife Conservation."

(Source: Pers. comm. S. E. Mott, October 26, 1987).

University of Nebraska State Museum

I. Stated missions:

"The State Museum shall be the depository of the University for specimens and related literature documenting the natural history and cultural heritage of Nebraska, the Great Plains, and whatever other areas are deemed suitable. Said specimens shall be maintained as a public trust and curated and preserved in an appropriate Museum division. These materials shall be made available for teaching; research, and interpretation, the results of which shall be communicated whenever possible to the scientific community and general public through publication, interpretive display, and educational programming. The University of Nebraska Museum of Natural History is a major state museum housed within a university administrative framework. Its missions, similar to those of the land grant University of Nebraska are teaching, research and service in approximately equal proportions as they relate to the collection areas of: anthropology, botany, entomology, parasitology, paleontology and zoology. There are no specific stated missions."

(Source: Letter from J. Janovy, Jr., 10 September 1987;
Bylaws 1.9.3 Univ. of Nebraska–Lincoln)

II. Relative importance of roles:

"Teaching, research and service in approximately equal proportions. Although, the Museum is in the process of attempting to upgrade and solidify its role as an educational resource for the public schools of Nebraska. The Museum feels its primary commitment is divided between the public and our related academic departments."

(Source: Letter from J. Janovy, Jr., September 10, 1987)

III. Organizational structure:
"The Museum is an administrative unit within the Office of Vice Chancellor for Research and Graduate Studies."

(Source: Same as above)

IV. Funding:
"The Museum is funded in part by direct state appropriations and independent income generating activities."

(Source: Same as above)

Nevada State Museum

I. Stated missions:
"The Nevada State Museum is a public service institution with statutory power to receive, collect, exchange, preserve, house, display and exhibit specimens and artifacts pertaining to the earth history, natural history, prehistory and history of the State of Nevada."

(Source: Statutory Authority NRS 381, 1939)

II. Relative importance of missions:
"Collection and collection development is the main emphasis. Their primary object is research and the results of which are published and serve as their main tool for education."

(Source: letter from L. DeVitta, 15 August 1985)

III. Organizational structure:
"The Nevada State Museum is a Unit of the State Department of Museums and History, a joint board of trustees. "

(Source: Statutory Authority NRS 381)

IV. Funding:
"The institution is funded through both State and private means, the State paying most of the salaries and the overhead while private funds are used for collections and exhibits and associated projects."

(Source: Letter from L. DeVitta, 15 August 1985)

New Jersey State Museum

I. Stated missions :
"The New Jersey State Museum has performed the traditional functions of collecting, exhibiting and interpreting in the broad areas of science, history and the arts."

(Source: New Jersey Administrative Code, 18A:73-1 to 18A:73-25)

II. Relative importance of roles:
"The museum's primary responsibility is to care for the state's collections while making them available for the public's enjoyment and education."

(Source: Same as above)

III. Organizational structure:
Formerly, the New Jersey State Museum was a unit within the State Department of Education. Since April 1983 the New Jersey State Museum has been a division of the New Jersey Department of State.

(Source: Same as above)

IV. Funding:

Funding for the Museum and its programs is through state appropriation, with additional income through memberships, contributions, and various fund raising projects.

(Source: 1983 New Jersey State Museum, Friends of the Museum annual publication)

New Mexico Museum of Natural History

I. Stated mission:

"The New Mexico Museum of Natural History, as mandated by law, is charged with increasing public understanding of New Mexico's aesthetic, cultural, natural, and scientific resources. The legislature of the State of New Mexico declared that the natural history resources of the state 'constitute a common heritage concerning which all persons should receive knowledge and benefit.' The purpose of the Natural History Act, therefore, was to create a state museum of natural history that shall 'collect, preserve, study and interpret materials representative of the natural history of the state and region and develop and maintain exhibits and programs of an educational nature for the benefit of the citizens of New Mexico and visitors to the state.' The institutional goal, as mandated by the Board of Trustees, is to develop as a superior regional museum dedicated to innovative education in the field of natural history."

(Source: Letter from L. Parker, January 5, 1988)

II. Relative importance of roles:

"There are four roles mandated of the museum by State Law: to collect, preserve, interpret, and study natural history specimens of New Mexico. The founding impetus for this institution was preservation of natural history specimens indigenous to this region. Collection activity is carried out through donations, volunteers, and some staff time. Research by the scientific staff is conducted both in the field and in six fully equipped laboratories with grants providing primary funding for staff research. As the museum has matured, the primary focus has changed from that of collection and preservation to one of education."

(Source: Same as above)

III. Organizational structure:

"The Museum is a division of the State Office of Cultural Affairs which is attached to the Department of Administrative Finance, a cabinet level of government. A Board of Trustees has fiduciary responsibility for the Museum and determines policy for the institution. The officer of Cultural Affairs maintains ultimate control over both director and Museum. The Board of Trustees is appointed by the governor and confirmed by the legislature for a term of four years."

(Source: Same as above)

IV. Funding:

"As a state-owned museum, 60% of the total operating budget is funded by the state legislature. The remaining 40% is generated from admissions, facility rentals, and donation from public and private sources."

(Source: Same as above)

New York State Museum

I. Stated missions:

"The New York State Museum is the primary state funded institution that researches, collects, preserves, interprets and exhibits the natural and human history of New York State.

As a unit of the State Education Department it has statewide responsibility to related cultural organizations. These responsibilities include the chartering of the

State's 1500 museum and cultural organizations and providing technical assistance, collection loans, cooperative exhibits and model museum education programs to the field."

(Source: Letter from M. Sullivan, 30 August 1985)

II. Relative importance of roles:

"The State Museum's functional priorities are research, collections, education, exhibits and service. This ordering places the research and collections as the building blocks on which the educational, exhibit and service programs are built."

(Source: Same as above)

III. Organizational structure:

"The State Museum is a department within the State Department of Education."

(Source: Same as above)

IV. Funding:

"Our entire budget is State appropriated through the Education Department."

(Source: Same as above).

North Carolina State Museum of Natural History

I. Stated missions:

"The North Carolina State Museum of Natural History is divided into the following sections which cooperate by program and maintain unique functions:

(a) educational services, which provides information on natural history, ecology and conservation;

(b) research and collections, which curates scientific collections and conducts research in natural history;

(c) publications which produces and distributes information;

(d) exhibits, which maintains exhibits on plants, animals and the physical world;

(e) faunal survey, which conducts biological surveys and resource inventories on vertebrate and invertebrate faunas; and

(f) H.H. Brimley Memorial Library, which serves as a natural science library and houses the archives of the State Museum of Natural History."

(Source: Statutory Authority G.S. 106-22(15))

II. Relative importance of missions:

"Research and collections are the primary focus of the Museum staff. However, education programs and exhibits are becoming more strongly represented. Ideally, both research (including faunal surveys and publications) and educational programs should have equal relative importance to the missions of the Museum."

(Source: Pers. comm. with Dr. J. Funderburg, August 1985.)

III. Organizational structure:

"The Museum of Natural History is a division of the North Carolina Department of Agriculture the result of an enactment of the 1879 State Constitution."

(Source: State of North Carolina Constitution, 1879).

IV. Funding:

"Funding for the Museum of Natural History is provided through the State Department of Agriculture."

(Source: Pers. comm. with Dr. J. Funderburg, August 1985)

Oklahoma Museum of Natural History

I. Stated missions:

"The mission of the Museum is to conduct research, participate in higher education, especially at the upper division undergraduate and graduate levels, disseminate information to the people of Oklahoma, and collect and preserve the tangible record of Oklahoma's natural and cultural history, which the Museum holds in trust for the people of Oklahoma. By its activities, the Museum preserves and interprets objects of cultural or scientific value, thus developing a greater understanding and appreciation of the natural and cultural heritage of the state, region, and world. The Museum has a responsibility to the University and the State of Oklahoma to prevent the loss or deterioration of its priceless collections through mismanagement, indiscriminate dispersal, or lack of proper curatorial care. The Museum functions as a research arm of the University, a service organization to numerous departments, a major educational bridge between the University and the people of Oklahoma, and the principal protector of Oklahoma's cultural and scientific heritage as reflected in tangible objects."

(Source: Executive Summary: P. Tirrell, April 5, 1988)

II. Relative importance of roles:
"The Museum fills a very complex role in its functions as the State Museum of Natural History an an Organized Research Unit of the University. The collections form a basic for the continuing reappraisal of man's knowledge of the past, present, and future. The Museum transmits knowledge about man and nature to the public by means of exhibits, interpretive programs, and other appropriate media through the Museum's Exhibits and Education Department. The Museum produces and disseminates scientific research through scholarly publications."

(Source: Same as above)

III. Organization structure:
"The Museum Director reports directly to the Universities Vice Provost for Research. The Director is the liaison between the Museum, the University, and the state."

(Source: Same as above)

IV. Funding:
"The parent organization, the University of Oklahoma, provides the major portion of the Museum's annual operating budget. This is supplemented by state and federal grants, contributions and earned income."

(Source: Same as above)

The State Museum of Pennsylvania

I. Stated mission:
"The State Museum of Pennsylvania collects, preserves, exhibits and interprets the history and material culture of the commonwealth. The museum is the principal repository for artifacts and collections relating to Pennsylvania's heritage. The State Museum conducts an ongoing program of acquisitions, care, conservation, research and education in Archaeology, Decorative Arts, Earth Science, Industry and Technology, Fine Arts, Military History and Natural History. Collections are presented to the public through permanent and changing exhibitions, educational programs, cooperative programs with other museums and public events."

(Source: Enabling legislation of 1905)

II. Relative importance of roles:
"The Museum has expanded its role beyond the initial concept of preservation to now include research, interpretive exhibitions and education. The State Museum utilizes its collections for each of these activities and for service to each of

its communities, the nation, the state, the local population. The State Museum operates with a primary goal of the preservation of artifact collections."

(Source: Same as above)

III. Organizational structure:

"The State Museum operates as an agency of the Commonwealth of Pennsylvania, with policy making authority vested in the Pennsylvania Historical and Museum Commission. The Commission executive staff provides over-sight for museum operations through fiscal and management staff. The State Museum Director has clearly defined responsivility for the implementation of policy as established by the Pennsylvania Historical and Museum Committee for the professional operation of the museum and for supervision of museum staff. The State Museum itself is organized in three operating divisions under the Director. Each division is headed by a Chief, a management level employee who works with the Director for planning, budgeting and supervision. The Director serves as liaison with the board and staff of the Friends of the State Museum, and with other units of the Commission which provide fiscal, personnel, design, marketing and other support."

(Source: Same as above)

IV. Funding:

"The museum is funded seventy four percent through state appropriations received through Pennsylvania Historical and Museum Commission. Five percent of funding is federal funds through competitive grant programs. Twenty one percent of funding comes from the non-profit, membership and support organization; The Friends of the State Museum.

(Source: Letter from C. Nold, September 3, 1987)

South Carolina State Museum

I. Stated mission:

"The South Carolina State Museum is a public, non-profit, educational institution, operated by the State of South Carolina, whose purpose is to collect, preserve, exhibit, interpret, and teach about the cultural history, natural history, arts, science and technology of the State in order to provide cultural enrichment, intellectual stimulation, learning, and enjoyment to the states' citizens and visitors. It is also the purpose of the State Museum to render assistance to other museums in the state."

(Source: Letter from T. Underwood, September 17, 1987)

II. Relative importance of roles:

"The Museum's roles in education, exhibition, collection, research and publication, and statewide services will be confined to the general purpose of the Museum. They will be unique to this museum and will not substantially overlap the roles of other institutions in the state. The Museum will represent and serve all the people of South Carolina.

The Museum's Educational role is to relate to the general public, teachers, and school groups through exhibits, publications, and special programs, the story of the cultural history, natural history, fine, decorative, and folk arts, and the scientific and technological resources of the State of South Carolina.

The Museum's exhibition role is to present the cultural history, natural history, arts, science and technology of the State of South Carolina in visual form, and to make information about these subjects accessible and understandable to as many people as possible.

One of the Museum's most important functions is to preserve significant cultural and scientific material related to South Carolina. To this end the Museum's collection role is to locate, acquire, and preserve in perpetuity a well-documented collection of cultural history, natural history, fine, decorative and folk arts, scientific and technological materials and artifacts pertinent to its other roles in education, exhibition, research and publication, and statewide services.

The museum will carry out research to document its permanent collections, to develop and continue its exhibition and education programs, to obtain information for publications to support its roles in education, exhibition, collection, and statewide services, and to conserve properly its cultural history, natural history, art, science and technology collections.

The Museum will lend assistance to other museums in South Carolina regarding all aspects of museum operations, including education, exhibition, collection, research and publication, and the conservation of cultural history, natural history, art, and science and technology collections."

(Source: Same as above)

III. Organization structure:
"The governing body of the museum is independent of and separate from existing commissions and departments of government. The South Carolina Museum Commission is composed of nine members appointed by the Governor. The primary function of the commission shall be the creation and operation of a state museum reflecting the history, fine arts and natural history, and the scientific and industrial resources of the state. Under the Commission's organizational structure, the director has delegated to the curators the responsibility for deciding which objects to acquire."

(Source: South Carolina State Museum Commission Annual Report 1985-1986)

IV. Funding:
"The State Museum's funded currently from two sources: public funds appropriated by the state legislature, and private funding. The primary funds are sought and administered by the state legislature, while addditional private funds are sought and administered by the South Carolina State Museum Foundation."

(Source: Letter from T. Underwood, September 17, 1987)

Texas Memorial Museum

I. Stated mission:
"The Texas Memorial Museum is an archival repository of the natural and social sciences dedicated to the collection and curation of specimens and data in these fields, research and graduate in structional study of the collections and to the dissemination of information about them through instruction, professional and popular publication, exhibits, and other public programs."

(Source: Annual Report, 1987)

II. Relative importance of roles:
"The Museum's role as an aid to research of the university is viewed as its primary role. Along with this, the management of the collection is ranked. The museum does provide educational and service to the community as a secondary role."

(Source: Same as above)

III. Organization structure:
"The Museum's Director is directly responsible to the University of Texas Vice President and Provost. The Director heads a staff of eight Curators of different disciplines, two laboratories of research and three museum departments (Publication, Exhibition, Administration)."

(Source: Same as above)

IV. Funding:

"The basic funding is provided by Texas legislative appropriation. This is supplemented by University of Texas research funds and museum generated funds."

(Source: Same as above)

Utah Museum of Natural History

I. Stated missions:

"The purpose of the Utah Museum of Natural History includes promoting the cultural and educational enhancement of Utah residents and tourists, plus supporting local and municipal museums throughout the State by collecting, maintaining and displaying the many objects which depict the past, present and continuing development of Utah's natural history."

(Source: Fact sheet, Utah Museum of Natural History, 1963)

II. Relative Importance of roles:

"The relative importance of our various roles has changed since our opening when emphasis was primarily on exhibits; education and development of and acquisition of collections (repositorial) and other services have taken a greater role as the museum has matured."

(Source: Letter from D. Hague, 22 July 1985).

III. Organizational structure:

"The Utah Museum of Natural History is located at the University of Utah and administered through the Vice President for Academic Affairs."

(Source: Fact sheet, Utah Museum of Natural History, 1985).

IV. Funding:

"Operating support of approximately 70% of costs is provided by the Utah State Legislature. Under this arrangement, the Museum is required to raise the difference of approximately 30% of its operating budget from admission charges, gift shop sales and donations."

*(Source: Fact sheet, Utah Museum of Natural History
and letter from D. Hague, 22 July 1985).*

Virginia Museum of Natural History

I. Stated Missions:

"To interpret Virginia's natural heritage within a global context, by methods and at levels relevant to all citizens of the Commonwealth.

'Interpret' means to conduct research as well as a broad range of educational programs, explaining and conveying the results of research done at the museum and elsewhere.

'Virginia's natural heritage within a global context' means that research and exhibits will be focused on Virginia, but with the understanding that within geology Virginia is part of an interconnected environment. Broad principles of natural history (such as evolution, plate tectonics) are to be represented in the exhibits as well as research programs.

'Relevant to all citizens of the Commonwealth' means that programs of research, education and exhibition will be balanced to include services and products of interest to scientists, educators and the general public. In order to maximize the use of our materials and human resources we will support branch museums as well as cooperative programs with other agencies and institutions."

*(Source: Virginia Museum of Natural History Policies
and Procedures Manual).*

II. Relative importance of roles:

"Exhibits and educational outreach service to citizens of the Commonwealth of Virginia are primary followed by research and collections focusing on Virginia within a global context."

(Source: Statement provided by D. K. Mathew in letter from P. Cato, 6 September 1990).

III. Organizational structure:

The Virginia Museum of Natural History is an independent state agency under the Secretary of Natural Resources. The museum is governed by the Board of Trustees that is appointed by the Governor, responsible for establishing priorities, and responsible to the state for the museum's finances. The Director answers to the Secretary of Natural Resources and Trustees.

(Source: Letter from P. Cato, 6 September 1990)

IV. Funding:

The museum received 97.6% of its funding from the State in 1989-1990 and the remainder through grants, membership, admission fees, programs, and store. This percentage will change as plans to diversify financially are implemented.

(Source: same as above).

Thomas Burke Memorial Washington State Museum

I. Stated missions:

"To increase our knowledge of the natural and cultural history of Washington State, the Pacific Northwest, and the Pacific Region by collecting, preserving, researching, interpreting, and exhibiting geological and zoological specimens and cultural artifacts. To achieve its purposes, the Burke Museum develops and preserves collections related to its area of interest; undertakes original scientific research and disseminates the results thereof; and educates at all levels in its areas of interest."

(Source: Washington State Museum statement of purpose, 1985).

II. Relative importance of roles:

"Because of our longtime university association, research is of paramount importance. . . . However, we are the major natural history museum in the Pacific Northwest and recognizing the obligations of our unique position we maintain a high public profile, much higher than many university museums. Guided tours, childrens programs, lecture series, workshops, field trips, and special exhibits are all regular offerings."

(Source: Letter from R. Auguszting, 26 July 1985)

III. Organizational structure:

"The museum is located on the campus of the University of Washington and is a department of the College of Arts and Sciences. The director reports to the Dean of Arts and Sciences."

(Source: Same as above)

IV. Funding:

"Funding is through the University of Washington. Although a state museum, state support accounts for less than half of the museum's revenue. The buildings, utilities, and some salaries are provided by the state. The rest of revenue comes from grants, contracts, gifts, and earnings from the coffee and gift shops."

(Source: Same as above)

Natural History Museums: Directions for Growth
Paisley S. Cato and Clyde Jones, editors
Texas Tech University Press, Lubbock, 1991, iv+252 pp.

NATURAL HISTORY MUSEUMS
AS A FUNCTION OF WILDLIFE MANAGEMENT

Catherine Carter Shropshire and R. Thomas Shropshire

Abstract.—Natural history museums that are governed by larger institutions must share parallel missions. The creation of the Mississippi Museum of Natural Science in 1935 as a division of the Mississippi Department of Wildlife, Fisheries and Parks paired two seemingly disparate philosophies. In 1985, the governing body of the Mississippi Department of Wildlife, Fisheries and Parks endorsed the development of a comprehensive planned management system for the department. The development of this system has created a unifying mission for the agency's divisions and has defined species oriented goals, objectives, problems, and strategies. Within this framework the museum's functions become solutions to problems facing the conservation of Mississippi's wildlife resources.

The Mississippi Department of Wildlife, Fisheries and Parks (MDWFD) was established in 1932. In the fall of that year, Frances (Fannye) A. Cook was appointed by the Commission as research assistant to the director of the department. Miss Cook brought to this position a background in teaching and a familiarity with museum functions. She recently had spent a year at the National Museum of Natural History conducting research and preparing specimens of plants and animals she had collected. During her early years with the Department, she conducted game censuses and surveys of freshwater fishes.

In 1935, Miss Cook planned a statewide plant and animal survey. This project was financed by the MDWFP in cooperation with the state colleges. During this project, thousands of specimens were collected and preserved. Eighteen regional museums were established and Miss Cook's dream of a state museum was realized when the Wildlife Museum was opened to the public as a permanent state museum on January 3, 1939. In 1971, the Wildlife Museum was designated by the legislature as the state's official natural science museum. The name was changed to the Mississippi Museum of Natural Science (MMNS) with the stated purpose, "To increase knowledge of the natural sciences particularly as they pertain to the conservation of Mississippi's wildlife resources, and to disseminate this knowledge among mankind."

During the past ten years, the museum's technical staff has grown and now includes five curators, three biologists with the Natural Heritage Program, a librarian, and an interpretation staff of five. The MMNS has focused its growing resources on research, collections, educational programming, and exhibits. The extension of the Museum's services to include the Nature Conservancy's Natural Heritage Program has increased museum responsibilities to include endangered species and critical habitat identification.

The museum's purpose and functions were often viewed by other department divisions as beyond the scope of a conventional game and fish commission. The department's perceived purpose was to manage wildlife resources for sportsmen. This concept reflects the main funding source for the department, which is sportsmen's license fees and federal funds derived from taxes on hunting and fishing equipment. Demonstrating that the sportsmen benefited directly from the museum's activities became an area of constant concern for the Museum in its project and budget development.

COMPREHENSIVE PLANNING

In 1985, the governing body of the Mississippi Department of Wildlife, Fisheries and Parks endorsed the development of a comprehensive planned management system for the department. This planned management system is composed of two parts: a strategic plan and an operational plan. The purpose of the planned management system is to (1) determine department priorities for managing specific wildlife resources, (2) determine where department efforts are most needed, and (3) determine where the money can be most productively spent.

Program Structure

The planning process will result in the following structure:
1. Strategic Plan (a five year plan)
   a. Department's mission statement
   b. Department's programs
   c. Program inventories
   d. Program goals
   e. Program objectives
   f. Program problems and strategies (solutions)
   g. Priority ranking of these programs, program problems, and program strategies
2. Operational Procedures Plan
   a. Writing and budgeting projects

    b. Reviewing and rating projects
    c. Evaluating project progress
    d. A cost accounting system
3. Annual Budget Development Procedure
    a. Writing and budgeting projects
    b. Reviewing and ranking projects
    c. Submitting to the legislature, the department's budget request, by project, program, or both, for annual funding consideration

Team Effort

A planning team composed of department heads has been organized to develop and implement this planned management system. As progress has been made, special task force teams, involving over 80 nondepartment individuals and approximately 300 of the 450 department employees, have provided specific information in a step by step process.

In the first step of this process, the planning team has developed the following mission for the department: "It is the mission of the Department of Wildlife Conservation to manage, conserve, develop and protect Mississippi's wildlife and marine resources and their habitats to provide continuing recreational, ecological, economic, edcational, aesthetic, scientific and social benefits for present and future generations." This mission statement, the first to be developed for the department in its 50 year history, obviously clarified the scope of the department's concerns and reflects the acceptance of the responsibility for all wildlife and the needs of the nonconsumptive as well as the consumptive user. A harmonious relationship between this mission and the museum's purpose became evident.

Following the development of the mission statement, the planning team identified the department's programs as follows: reservoir, lake and oxbow fish; rabbits; river and stream fish; squirrels; small impoundment fish; bullfrogs; nongame and endangered resources; furbearers; MDWFP lake fish; turkey; gallinules, rails and snipe; alligator; woodcock; blue crab; dove; oyster; white-tailed deer; shrimp; ducks and coots; saltwater fin fish; wood duck; hunter education; goose; boating; bobwhite; and waterfront. Basically, these programs reflect the monetary and personnel commitment currently being placed on these areas of concern. This list will be reviewed and revised periodically to adapt to the changing needs of the wildlife resources in the state.

Task force teams were assigned to address each of these programs. The participants on these teams included department personnel from all technical and support functions, conservation officers (law enforcement officers), and representatives of universities, the U.S. Forest Service, U.S. Army Corps of Engineers, U.S. Fish and Wildlife Service, and interested or knowledgeable citizens. Using a facilitiated meeting format to obtain concensus decisions, these teams developed management histories of the resources, supply and demand analysis, goals, measurable objectives, problems, and solutions for each of these programs.

Developing a Strategic Plan

For purposes of museum planning, the nongame and endangered resources strategic plan will illustrate how the process worked and how the museum will fit into the program. This team consisted of biologists with our Bureau of Marine Resources, Department of Wildlife and Fisheries biologists, wildlife area managers, museum curators, heritage program biologists, the museum librarian, private individuals, and a U.S. Fish and Wildlife Service biologist.

The team first met and assigned a person to write a history of the nongame and endangered resources in the state. These histories provide overviews of (1) where, when, and why the department started managing the resource or supplying the service, and (2) the current status of resource management activities and services. The history was reviewed and edited by the team and tailored to conform to the one-page format for the strategic plan. The history then became the basis for determining supply and demand analysis.

For programs involved with consumptive use, the supply and demand analysis incorporated harvest data including hunter and trapper survey statistics to provide some general assumptions about use of the resource. For nongame and endangered resources, we referred to the 1980 National Survey of Fishing, Hunting and Wildlife Associated Recreation (1982), which estimated that 570,000 Mississippians participated in some form of nonconsumptive use of wildlife. Also the passage of the Mississippi Nongame and Endangered Species Act, The Wildlife Heritage Act, The Natural Heritage Act, and The Mississippi

Wildlife Heritage Fund Checkoff Law were mentioned as indicating a concern for wildlife resources in the state.

The team next developed the program goal: "It is the goal of the Mississippi Department of Wildlife, Fisheries and Parks to insure the perpetuation of nongame and endangered species, communities and ecosystems in their natural state." The development of the objective for the program followed. "It is the objective of the MDWFP to manage nongame species to prevent their becoming endangered and to enhance the recovery of endangered species so that they may be delisted."

Problems that will be encountered in reaching that objective were prioritized as follows: (1) lack of funding to support nongame and endangered resource management; (2) lack of baseline scientific knowledge about species, communities, and ecosystems to establish management strategies; (3) lack of education and public awareness about nongame and endangered resources; and (4) loss of habitat.

At this point the goal, objective, and problems may not resemble a typical museum long-range plan; however, when strategies were discussed, the museum's functions became solutions to the problems and revealed a common focus for our energies and resources. The strategies or solutions which came out of the next session were prioritized as follows:

1. Problem: Lack of funding—strategies
   a. Increase public and MDWFP employee awareness of this resource
   b. Identify revenue sources (i.e., user fee system on public lands, tax on equipment used by nonconsumptive and consumptive users)
   c. Increase political awareness (i.e., to purchase habitat)
2. Problem: Lack of baseline scientific knowledge—strategies
   a. Conduct inventories of species, communities, and ecosystems
   b. Conduct status surveys
   c. Conduct life history studies
   d. Develop a computer data base of information that is easily accessible
   e. Set priorities for nongame species and develop management plans for utilization of those species for both consumptive and nonconsumptive users
   f. Publish results of research
   g. Conduct literature searches

   h. Set priorities for research and recovery efforts
   i. Designate indicator species
3. Problem: Lack of education and public awareness—
   strategies
   a. Produce films and videos
   b. Expand MMNS educational programming
   c. Expand educator workshops
   d. Establish a natural science core curriculum in the
      school system
   e. Publish research in popular form
   f. Establish nature centers
   g. Expand MMNS permanent and traveling exhibit
      projects
   h. Expand MDWFP (Hunter Education) staff to conduct
      projects for schools and support organizations
   i. Cooperate with environmental groups
   j. Identify target groups with similar goals and objectives
   k. Establish hiking trails
4. Problem: Loss of habitat—strategies
   a. Establish a wildlife legacy trust fund and purchase
      habitat
   b. Support new legislation such as the stream bill and
      land zoning to prevent further destruction of habitat
   c. Provide economic incentives for landowner (i.e., con-
      servation easement, conservation a habitat protection
      by private landowner)
   d. Regulate and enforce current laws to prevent further
      destruction of habitat
   e. Develop methods to manage acquired land for non-
      game and endangered resources
   f. Educate public about the effects of habitat destruction
   g. Cooperate with other state and federal agencies
   h. Relate habitat to the quality of human life
   i. Identify critical habitat

   In all other team meetings the recurrent theme of lack of
public awareness and education was discussed and the
museum's resources noted as as important means of attacking
the problems.

Project Funding

Projects will be written to address a solution or problem listed in a particular program. These project proposals will outline the personnel and money required for the project much as a grant proposal is written. These proposals will be reviewed and ranked through the chain of command and funded in order of priority. Basic operations and maintenance budgets will receive priority funding from the appropriated funds assuring that all divisions retain the fundamental support they require.

CONCLUSION

The comprehensive planning process has demonstrated that the MMNS's functions are both complimentary and supportive of the Department of Wildlife Conservation's mission and goals, and has created a more open working relationship among divisions. As the planning process develops and specific project proposals are written, this cooperative effort should provide better data, educational services, and overall understanding of Mississippi's wildlife resources. The Mississippi Museum of Natural Science is in a unique situation in the nation with regard to its relationship with a department of wildlife conservation, housing the Natural Heritage database, and being accredited by the American Association of Museums. The potential for producing a positive impact for Mississippi's wildlife resources is unlimited and we are confident that the mechanism is being developed that will allow us to reach that potential. The development of a comprehensive planned management system has created a unifying mission for the Department's divisions and has defined species oriented goals, objectives, problems, and strategies which will require the combined efforts of all divisions to realize.

LITERATURE CITED

U.S. Department of the Interior, Fish and Wildlife Service and U.S. Department of Commerce, Bureau of Census. 1982. 1980 national survey of fishing, hunting, and wildlife associated recreation. U.S. Govt. Printing Office, Washington, D.C., 156 pp.

Collections

Natural History Museums: Directions for Growth
Paisley S. Cato and Clyde Jones, editors
Texas Tech University Press, Lubbock, 1991, iv+252 pp.

THE CONSERVATION OF NATURAL HISTORY COLLECTIONS: ADDRESSING PRESERVATION CONCERNS AND MAINTAINING THE INTEGRITY OF RESEARCH SPECIMENS

Carolyn L. Rose

Abstract.—Conservation involves three basic areas of activity: preventive care, treatment, and research. In recent years, a greater emphasis has been placed on preventive activities, both as a realistic approach to the conservation of extensive museum holdings, and in response to increasingly sophisticated analytical use of collections which may be affected by treatment.

Systematic preventive care is achieved through a long range conservation plan that incorporates preservation policies and procedures into the daily activities of the museum, and develops strategies for addressing conservation problems in a logical, feasible manner.

Although this approach is especially valid for natural history museums, field collections generally require some type of stabilization treatment prior to storage and preventive care. However, commonly practiced preparation techniques are frequently based on traditional methods developed a century or more ago, rather than on modern scientific knowledge, and do not necessarily complement the current research use of specimens.

The scientific community must recognize the need for a professional conservation approach to treatment and documentation, and the need for preservation research. A commitment to collections care should be implicit in all collecting activities, and is required to obtain the support and funding necessary to ensure the future and integrity of scientific specimens.

Conservation is not a new term to those involved with the natural sciences, and in this context, conservation usually refers to steps taken to ensure the preservation of natural resources through preventive means. Conservation within the context of museums, on the other hand, is generally associated with the treatment of objects, most often with the restoration of works of art. During the last decade, however, this traditional, intervention-oriented approach to museum conservation has shifted in focus to that of prevention, especially for large collections. To those unaware of this current direction, museum conservation still connotes treatment, and the applicability of this field to natural science collections seems minor, and relevant information is often ignored.

In addition to preventive care and treatment, research in conservation is an equally important activity of this discipline. Current museum conservation research, in fact, often probes into the very problems causing deterioration of natural science collections and their accompanying documentation, such as labels and field notes. Such research also may clarify other disturbing problems associated with scientific inquiry if explored more fully. Many current studies focus on the nature of materials, their interaction with the surrounding environment, and the mechanisms of material alteration and deterioration. These materials studies are fundamental to determining methods of preventing deterioration and maintaining the physical and chemical research potential of museum specimens.

Of the conservation information that now exists, most of the preventive practices and some treatment techniques are directly transferrable to the natural sciences. In some instances experimentation is required to develop treatment modifications to meet the specific preservation goals of these collections. In other cases, specialized research is needed to address totally different kinds of problems, such as those associated with wet collections.

One of the major problems plaguing the preservation of natural history specimens is that the transfer of this information often occurs on an ad hoc basis in order to solve specific, immediate problems: formalin solutions are becoming too acidic, label inks are running, shells are exfoliating, and skeletal materials are powdering. Too often the solutions to these problems reflect an expedient and simplistic approach: add buffer to the formalin, first rinse the labels in alcohol, or coat the surfaces of the shells and bones. Although this approach can answer an immediate problem, it does not usually address the causes of the problem or the long-term solution. In some cases, these methods may even exacerbate the situation.

Adopting empirical, nonscientifically based methods is not the answer to problems in the preservation of natural history collections, nor does the answer lie totally in the direct transfer of techniques from other disciplines. To appropriately address the preservation problems of these collections and those of the future, and to maintain the research integrity of scientific specimens for years to come, a new discipline within the museum conservation field must be developed for the natural sciences, just as it was developed for anthropological collections only a few years ago (NIC, 1984).

Most problems affecting the preservation of natural history collections originate from field and laboratory preparation techniques, which I will discuss later, and the museum environment.

Upon mentioning the museum environment, a typical reaction, is that one quickly defends the reasons why proper relative humidity and temperature levels cannot be achieved in their museum. The museum environment, however, goes beyond the general museum climate and includes not only the specific microenvironment that surrounds each specimen, but also the conditions to which the specimens are subjected. This includes fumigation as well as the use of specimens in exhibitions and research. For example, it is common practice to employ pest eradication methods on all botanical specimens that enter the museum, whether they require it or not. Furthermore, someone in the borrowing museum usually decides which methods will be employed on the specimens rather than the lending museum.

The policies, procedures, and guidelines employed in museums all contribute to the make-up of a specimen's environment. For this reason, a study of the total picture is required to determine the best and least intrusive preservation methods for collections.

Many grants programs have recognize the validity of this total environmental approach, and have restructured their guidelines to fund projects that would produce the most lasting impact on collections care. The Institute for Museum Services (IMS), for example, requires museums to complete a conservation survey and develop long-range preservation plans, before funding for treatment is requested (Institute for Museum Services, 1989). This is not to say that specimen preparation and object treatments are not important steps in preservation, but rather that provision for proper environmental controls, including sound policies and practices, should be the first step in the preservation process.

To facilitate the survey process, a new Conservation Assessment Program (CAP) is being funded by IMS. This program, administered through the National Institute for Conservation (NIC) provides free conservation assessments to museums on a first come first serve basis. (Institute of Museum Services, 1990)

THE CONSERVATION ASSESSMENT PROCESS

The goal of a conservation assessment, or survey, is to provide information necessary to develop a collections care and maintenance plan (Appelbaum and Himmelstein, 1986; Virginia Association of Museums, 1987; Shepard, 1988; National

Institute for Conservation and Getty Conservation Institute, 1990). This plan details the steps necessary to achieve specific preservation objectives and proposes methods of implementation. Critical elements in the design of a collections care strategy are (1) that the objectives set forth be realistic, and (2) that the collections care plan be integrated into the overall long-range plans of the museum. The survey process begins with a review of the mandate of the museum (the purpose in collecting and preserving specimens) and an inquiry into how the specimens are used at present and how they will be used in the future. The awareness of those ultimately responsible for the care of these collections is an important factor as well. For example, are museum trustees aware not only of their ethical, but also of their legal responsibilities for the collections that they hold in trust? Are they aware of the steps necessary to attain proper care for the collections?

An analysis of the staff also, in terms of their collections responsibilities, is integral to developing this plan. Who is responsible for the care of collections? Have these collections care responsibilities clearly been defined in position descriptions? And most importantly, do those assigned this responsibility have the knowledge and skills necessary to perform these tasks?

Most museums recognize the need to define their collections management policies in terms of specimen collection and documentation requirements. More frequently than not, these same collection policies do not clearly address collections care, except as it pertains to insurance requirements. In some cases, special rules and procedures have been developed for type collections, although enforcement of these regulations often is applied inconsistently. Because the attitudes of museum staff members, as well as both written and unwritten policies and procedures pertaining to the use and care of collections are among the most important factors influencing collections care, discussions with as many museum personnel as possible are an important part of the survey process.

In addition to the uses of collections, other aspects of their physical and chemical environment must be considered. A systematic approach to studying these factors usually involves an examination of the environment from the macro- to micro-level. The assessment often begins outside the museum with questions concerning the local climate and an examination of the physical fabric and condition of the building. The climate of the interior subdivisions as well as the intimate storage and exhibition environment should be studied. At each level, the ambient conditions of the larger can impact on the conditions of the next smaller. For instance, precautions taken at the structural level, such as proper building seals and the location of eating facilities (in combination with effective policies concerning the movement of food and specimens), can eliminate most pest infestations and hence the need for chemical treatment. In terms of the microenvironment, it should be remembered that exhibition and storage-case construction materials, boxes, fabrics, labels and mounts, as well as adjacent museum specimens, can accelerate the deterioration of the collections through physical and chemical changes.

All of these factors—a museum's mandate, staffing, policies and procedures, use of collections, and the macro- and microenvironment for storage and exhibition—must be considered to evaluate the needs of the museum collections and to formulate recommendations that can be used to develop long-range conservation plans. To be effective, such an assessment must be conducted by trained professionals in cooperation with those staff members who would formulate goals and implement long-range plans. The Getty Conservation Institute and the NIC recently have developed a handbook containing guidelines and checklists which can be used as a tool in conducting a conservation assessment in these areas. (National Institute for Conservation and Gerry Conservation Institute, 1990.)

THE COLLECTIONS CARE PLAN

A collections care plan can be effective only if it is realistic, if it is intimately linked with other museum objectives, and if collections care is, in fact, a museum goal. Conservation priorities for museums should be organized on short-term, intermediate, and long-term bases, in concert with other museum planning and fund-raising activities. Although grants are available to achieve some of the more difficult long-term goals, conservation, or collections care, should be a line item in the museum operating budget. Approaching the funding of these activities solely on the basis of soft monies is a delusive and ineffective approach.

Many immediate and short-term goals can be accomplished within a museum at minimal cost. These include establishing policies and procedures to improve the care of collections, and incorporating them into collections management documents and policy statements. Upgraded documentation procedures that include information on specimen condition and preservation treatments employed are such an example. These are fundamental requirements for professional preservation practices that should be met. In addition, information on standards and preventive practices is available through training courses in conservation awareness and collections care, and new forums designed to share expertise, such as the annual meeting of the Society for the Preservation of Natural History Collections (SPNHC). The provision of funds to attend such programs, to increase conservation knowledge and skills, is a responsibility of museum management that should not be ignored, especially because monies for such education and training of staff are readily available through several grants programs.

There are workable solutions to solving collections care needs within each museum. Recognizing that needed changes will occur over the long-term, breaking down goals into manageable objectives, and planning for collections care in a logical, systematic manner is key to being able to achieve proper care.

SPECIMEN PREPARATION TECHNIQUES AND CONSERVATION RESEARCH

For natural science specimens, preventive museum practices generally begin after field and laboratory preparation techniques have been completed. These specimen preparation techniques are conservation treatments, as are fumigation and other practices that result in physical and chemical changes in the specimen. Advances in treatment techniques for natural science specimens, however, have not kept pace with advances in treatments for other museum objects, even though relevant information is available. In fact, certain treatments currently employed in natural science collections may be among the primary causes for deterioration of many specimens. The preliminary findings of the Carnegie Museum's survey of pre-1900 mammal type specimens, for example, indicate that a collector's preparation technique is more often the factor that determines the overall stability of specimens than is the storage environment (Hawks, 1988, pers. comm.). This statement does not contradict previous statements concerning the importance of controlling the environment, but rather reflects the fact that the environments at most museums surveyed were generally the same.

If variances in preparation techniques are known to be a major factor in determining long-term preservation, why then, are the same techniques being perpetuated decade after decade with only slight modifications? Why is there a dearth of literature on sound, scientific research results for preservation techniques, that are comparable to those for other scientific studies? Some might blame this problem on funding sources for not supporting preservation research. But is it not, in fact, those leaders of the scientific discipline who set the priorities? Moreover, it is obvious that the development of better specimen preservation methods is not a museum priority when the technical staff is recognized only for their contributions to scientific research.

New initiatives in preservation research for natural science specimens are needed to elevate collections care activities to the sophistication of that in other scientific studies and to a level comparable with that of current conservation research in other museum disciplines. Specimens that are needed for comparative research and education have been lost because of poor preservation techniques. This loss will continue if improved techniques ae not developed. Furthermore, current preparation and fumigation treatments may negate the ability of scientists to conduct certain kinds of biochemical studies in the future. The development of new preparation techniques, based upon the application of modern materials, will take years to accomplish. Even slower to occur will be the changes in attitude and philosophy necessary to promote these new techniques.

In the interim, there are two activities that should be pursued by those responsible for the care of collections and by those in a position to influence policy making. First is the development of procedures for the documentation not only of the condition of specimens, but more importantly, of all chemicals that are used on or come in contact with specimens. One of the most troubling aspects of evaluating the long-term effects of various treatments on the preservation of specimens is that records pertaining to treatment techniques are, at best, incomplete and imprecise. The Conservation Committee of SPNHC has developed guidelines for documenting the condition of specimens and treatment reports, for geological, paleontological, and biological collections. (Fitzgerald, 1988; Garrett, 1989) It is the responsibility of those charged with the care of collections to adapt guidelines such as these to their museum situation and to advocate their use not only for type specimens, but for all catalogued specimens.

A second and equally important activity is the promotion and advocation of the importance of specimen treatment research to museum administrators, to professional associations and societies, and to funders. Unless the scientific community is willing not only to recognize the importance and validity of conservation and preservation research, but to assume the responsibility for promoting this new field, little advancement will occur. Included in this, of course, is internal recognition, encouragement and support for those pursuing these endeavors within one's own museum.

SUMMARY

As our environment changes and more and more species become extinct, the fervor to collect has intensified. As new questions are asked and new discoveries are made, the need to recheck previous findings and test new methodologies and hypotheses has increased as well. Important also is the need to share these new findings and to increase knowledge through exhibitions and educational programs. All of these activities are related to natural history museums.

Equally important, although frequently neglected in this process, is providing proper attention to improving the care and preservation of these specimens. Natural science museums, by the nature of their fiduciary responsibilities, are morally and legally obligated to protect the specimens that they house and collect. To perform this function, museums must develop logical and feasible conservation strategies that are integrated into the overall museum plans.

A basic museum conservation philosophy that includes regular on-going support for conservation activities, recognizes and rewards advancements in this field, and encourages the pursuit of conservation and preservation education and research is needed. This philosophy and direction must be advanced by professional scientific organizations and journals, and its importance must be realized by funding organizations and by educational institutions where new scientists are being trained.

Only through concerted efforts on a variety of fronts can we advance the field of natural science conservation to the sophisticated level of other museum disciplines and to the level required to insure the preservation of the collections for future generations.

Literature Cited

Appelbaum, B. and P.R. Himmelstein. 1986. Planning for a conservation survey. Museum News, 64:(3):5-14.

Fitzgerald, G. R. 1988. Documentation guidelines for the preparation and conservation of paleontological and geological specimens. Collection Forum, 4(2):38-45.

Garrett, K. 1989. Documentation guidelines for the preparation of biological speciments. Collection Forum, 5(2):47-51.

Hawks, C. A., and C. L. Rose. 1987. A preliminary list of conservation resources for the care of natural history collections. Conservation Committee, Society for the Preservation of Natural History Collections, 24 pp.

Institute of Museum Services. 1990. 1991 Conservation Assessment Program: grant application and information. IMS, Washington, D. C., 20 pp.

Institute of Museum Services. 1989. 1990 Conservation project support: grant application and information. IMS, Washington, D.C., 79 pp.

National Institute for the Conservation of Cultural Property. 1984. Ethnographic and archaeological conservation in the United States. NIC, Washington, D.C., 20 pp.

National Institute for the Conservation of Cultural Property and the Getty Conservation Institute. 1990. The conservation assessment: A tool for planning, implementing, and fundraising, NIC, Washington, D.C., 50 pp.

Shepard, L. 1988. Shaping a conservation plan through general surveys. Newsletter of the American Institute for Conservation, 13:1-2.

Virginia Association of Museums. 1987. Conservation awareness technical assistance leaflet. VAM, Richmond, Virginia, 9 pp.

Natural History Museums: Directions for Growth
Paisley S. Cato and Clyde Jones, editors
Texas Tech University Press, Lubbock, 1991, iv+252 pp.

COLLECTIONS CARE
IN A SMALL NATURAL HISTORY MUSEUM

Valeen Silvy and Paisley S. Cato

Abstract.—Using the Brazos Valley Museum as a case study, this paper explores
the policies, procedures, and resources that may be used for collections care in
small museums. The Brazos Valley Museum was started in 1962 with a majority of
the specimens it currently maintains. Little effort had been made until 1982 to
increase the size of the collection or properly care for the specimens. Additions
have been made in recent years, through private donations, primarily for exhibition purposes.

During the past seven years, collections care has become a viable, although not
primary, function of the museum. First, written policies and procedures were
developed through the use of local talent and expertise. Next, an in-house survey
of specimens and storage conditions was made. The results of this survey were
used to develop a long range plan for collections care. To implement the plan and
acquire materials, financial support, and manpower, two important techniques
were used: (1) begging, borrowing, and scrounging, and (2) grant writing for both
state and federal museum support programs. Although not complete, there is an
on-going effort to improve the level of collections care.

Since 1982, the Brazos Valley Museum (BVM) in Bryan,
Texas, has taken steps to develop a methodical and careful
process to ensure the preservation of our natural history collections. Our museum's collections may be similar to many
other small museum collections, comprised of a mixture of
random donations, specimens that intentionally have been
acquired, and those that came with large gifts. For example,
many of our bird and mammal mounted specimens are from
a private donation when a local "Frontier Town" closed. Our
primary emphasis is on zoological, geological, and paleontological materials, approximately 4500 specimens altogether.
A collection of even this size can strain the resources of a
small museum when collection care has not been an integral
function of the museum.

The goal for our collections care project is not only to care
properly for the existing collections, but to demonstrate that
such care should be viewed as worthwhile by the general
public. The latter enables us to receive larger, more select
donations and to solicit more successfully specific specimens
to aid the institution's purpose.

To understand how our current collections care philosophy
is a complete reverse from our earlier years, it might help to

know a little of the museum's history. The BVM was started
in 1962 as a children's museum and was run entirely by volunteers. It is located in a small urban area of approximately
110,000 people, an area that also houses a major university
with almost 40,000 students, Texas A&M University.

The primary emphasis of the museum was originally only
on educational programming for young children. Since that
time, we have expanded our philosophy and programs to
serve families and adults in the community. The primary
goal of the BVM is to interpret the natural history of the
Brazos Valley, a region located between Dallas and Houston,
through its collections, exhibits, and programs. Art, history,
and cultural activities are incorporated into this purpose
when they serve to further interpret the natural environment
of the local area. In addition to our expanded philosophy
and programming, we have evolved into a professionally
trained staff and a board of trustees that recognizes the need
to maintain professional museum standards. The museum
currently employs three full-time museum professionals,
three part-time support positions (with assistance from the
Senior Texas Employment Program), and four to six full-time
positions for the summer nature camp. Our operating
budget is approximately $100,000 annually.

RATIONALE FOR A COLLECTIONS CARE PROGRAM

Weil (1986) stated, "museums come in a variety of types
and sizes, collect to different degrees and in different ways,
provide different services and have different needs." Yet the
existence of collections in a museum dictates certain legal
and ethical obligations on the part of the museum staff and
board of trustees; it is to the museum's advantage to develop
policies and procedures to ensure the care of the collection
items in a perspective that suits the needs and resources of
the museum, regardless of its size.

One of the primary reasons a small museum needs to
develop a collections care program is the issue of liability.
Ethically and legally, the staff and trustees of a museum are
liable for the well-being of specimens held in the museum's
care (American Association of Museums, 1978; Malaro,
1985). These standards apply to the small museum just as
much as larger research museums. And in many ways, the
smaller museum may be held to closer public scrutiny merely

because of its smaller size and smaller audience. There is a much better chance of its public having first-hand knowledge of the museum's successes and failures. These public members see the deterioration as they view specimens on exhibit or request to see the specimen that was on exhibit last year. They also remember what their friends gave to the museum 20 years ago and want to know where it is, why it was sold, and so on. This public frequently becomes more intimate and possessive of the museum's collection. Any legal action against the museum claiming negligence, or worse, could put as big a dent in the museum's budget as it would for a larger institution, if, that is, the museum could afford to defend itself at all.

A second, and very important, reason for developing a collections care program in a small museum is to guide the growth and direction of the museum. A collections program can help the museum carry out its purpose or mission statement. For example, the purpose of the BVM is to interpret the natural environment of our region for the community, but we wish to do so in such a way as to avoid duplicating services available through other organizations in the community. Our neighbor, Texas A&M University, has a number of nationally prominent research collections, including the Texas Cooperative Wildlife Collection, one of the largest and most comprehensive collections of vertebrate specimens in the southwestern United States. There is little demand or need to attempt to duplicate a research collection for the local area, as researchers generally can gain access to the university's collections. However, the public in the local community does not have ready access to those collections. Therefore, the BVM collections policy is directed to the care and preservation of specimens that can be used by the local public for reference, exhibits, and teaching. In view of this, our acquisition policy and management procedures take into account a division of collection use. Some specimens are set aside for study purposes, to be handled only by trained individuals, whereas others can be used with varying degrees of security and exposure to the environment.

DEVELOPING A COLLECTIONS CARE PROGRAM

Philosophy.—Probably the least expensive, but one of the most critical steps in developing a collections care program at the BVM, was to accept fully the philosophy that we needed to care for the collections in an active rather than passive manner. It might be said that small is only a frame of mind, but in view of the demands placed on a small staff with limited resources, it was too easy to push collections care to the back burner with excuses that we just didn't have the time and resources to handle them according to big-institution standards. However, once we accepted that we needed to attack the collections problems on a scale that suited our institution, we realized we could accomplish a great deal.

Written documents.—The second phase of our process was to develop written documents to guide the collections program. These documents would serve not only as guidance for the staff, as they do in all museums, but would alert the Board of Trustees of their responsibilities to the collections. As so little had been done previously to the collections, this area of responsibility was new to most trustees. And because the trustees are indeed liable for any possible legal aspects of mismanagement of the collections, this process has been an enlightening experience for all.

During 1982–1983, a collections care specialist (Cato) from Texas A&M University surveyed and organized the collections to assess the extent of the collection, develop a basic specimen inventory and determine basic identification and registration needs. With this information, in 1984, the Collections Committee of the Board of Trustees developed a written policy statement on collections. This committee included Cato, an attorney, two of the university's professors and the museum's director. The policy, approved by the full Board of Trustees, has provided the framework for the remainder of our collections care program (and is available upon request from the Brazos Valley Museum).

The next documents to be written were procedural documents, and these are still being developed. Admittedly, progress is slow due to the high level of demands on each staff member. However, in attempting to develop procedures, we tried to set ideals, and of course, modified these to fit the reality of the small museum. The procedures initially included developing a proper paper trail for each specimen following references such as Williams (1977) and Cato (1986). This paper trail needed to be done for our existing collection as well as any new specimens coming into the museum. Eventually our written procedures will include complete pest management plans, disaster plans, and so on—but all fitted to the needs and resources of our institution.

While developing these written procedures, it became clear that the museum required outside assistance before it could continue to mature. In 1984, we requested a Museum Assessment Program I, administered by the American Association of Museums, and received it. Then, we began an in-house assessment, using university personnel for expertise, to review the museum's environmental conditions, storage conditions, specimen conditions, and compatibility with existing use of the facility and specimens. This process identified the major problems relative to collection care and resulted in a series of written recommendations concerning improvement of registration, cataloging, and storage needs. In addition, we developed a written five-year plan for improving the collection storage environment, and the management and conservation of specimens. This time frame may not always be on target and there are many objectives remaining to be accomplished, but it has allowed us to recognize our priorities. Recent articles by Beale (1987), Hutchins (1987), and Rose (this volume) discuss the value of general conservation surveys and long-range conservation plans. In 1989, the BVM obtained an Institute of Museum Services Conservation Project grant for a general conservation assessment of the museum's facilities and collections. The museum will use the result of this assessment to evaluate and revise the long-range conservation plan.

IMPLEMENTATION OF THE COLLECTIONS CARE PROGRAM

The final consideration in our Collections Care Program involved recognition of our limitations and a dedicated effort to resolve them. First, no one on the staff was qualified to care properly for all the collections. Even in a university town, it is difficult to find someone who has the expertise to do five jobs, the normal number required of any one staff person in a small museum. Secondly, the museum did not and still does not have the financial resources to purchase the necessary equipment, or hire full-time collection staff. So, we turned to volunteers, to seeking donations of equipment, and to grant-writing to rectify our situation.

For expertise in collection care, we frequently have turned to the personnel of the university collections. These individuals have been helpful in a variety of ways: identifying and cataloging specimens; providing students to work with the museum's collections; recommending knowledgable volunteers; training our staff and volunteers on proper collection documentation, handling and storage; facilitating donations of equipment from the university surplus to the museum; and writing grants to support the museum's functions. The donated equipment has included steel map cases and handmade vertebrate specimen cases. With monies from both federal (Institute of Museum Services Conservation Project) and state grants (Texas Historical Commission), we have been able to purchase paleontology and vertebrate specimen cases and closed cabinets for fluid-preserved specimens, as well as for mounted specimens.

Just as important as our scrounging and grant-writing, however, has been the decision to commit museum dollars—even a few—to care for the collections. The museum cannot afford a full-time collections manager at this time; instead, priorities among our full-time positions were reallocated, so that at least part of one position is responsible for managing the collection. The board of trustees acknowledged the need for trained personnel in collection management by approving funds to send the director to the Collection Care Pilot Training Program for Natural History Collections.

In summary, we set a goal to properly care for our collections, developed written policies to guide the staff and board, and with borrowed staff and expertise, developed a long-range plan for improving the conservation and management of our collections. As we do not have all the resources we need in-house, we have involved the community in assisting us in realizing our goal.

LITERATURE CITED

American Association of Museums. 1978. Museum ethics. American Association of Museums. Washington, D.C., 31 pp.

Beale, A. 1987. Long range conservation planning for museums. Pp. 1-28, *in* Conservation Awareness Technical Assistance Leaflet, Virginia Association of Museums, Richmond, VA, 9 pp.

Cato, P. S. 1986. Guidelines for managing bird collections. Museology, 7:1-78. The Museum, Texas Tech University.

Hutchins, J. 1987. Conservation surveys. Pp. 5-6, *in* Conservation Awareness Technical Assistance Leaflet, Virginia Association of Museums, Richmond, VA, 9 pp.

Malaro, M. C. 1985. A legal primer on managing museum collections. Smithsonian Institution Press. Washington, D. C., 351 pp.

Rose, C. L. 1989. The conservation of natural history collections: Addressing preservation concerns and maintaining the integrity of research specimens. pp. 00 *in* Natural History Museums: Directions for Growth. (P. S. Cato and C. Jones, eds.). Texas Tech University Press, Lubbock, 00 pp.

Weil, S. 1986. Questioning some premises. Museum News, 64:20-27.

Williams, S. L., R. Laubach, and H. H. Genoways. 1977. A guide to the management of Recent mammal collections. Spec. Pub., Carnegie Mus. Nat. Hist., 4:1-105.

Natural History Museums: Directions for Growth
Paisley S. Cato and Clyde Jones, editors
Texas Tech University Press, Lubbock, 1991, iv+252 pp.

CONSERVATION PROBLEMS OF FLUID-PRESERVED COLLECTIONS

John E. Simmons

Abstract.—Preservation in alcohol has been practiced only since the mid-1600s. Formaldehyde did not come into use as a fixative until the 1890s. The techniques currently in use with fluid collections are traditional methods discovered by trial-and-error, rather than developed by rigorous testing. Specimen treatment in the field and in the museum is inadequately documented. Most fluid-preserved specimens are fixed with formaldehyde, which may result in acidification of the solution and damage to the specimens. Many kinds of tags, labels, closures, and jar gaskets in use in collections react with the preservative and can damage specimens. Specimens should be shielded from ultraviolet light, not be allowed to dehydrate, and kept under cool conditions, avoiding fluctuations in relative humidity and temperature. Research is needed to determine the best fixatives and preservatives for fluid collections. All traditional procedures used with fluid collections should be rigorously tested. A conservation code of ethics should be adopted for natural history collections.

HISTORICAL PERSPECTIVE

The earliest known attempts by human beings to preserve plant and animal specimens from deterioration do not have their origin in scientific practice, but rather in magic (Bedini, 1965). Preservation is thought to have begun with animals or animal parts saved as objects for religious or magical purposes (James, 1957).

The earliest known form of successful preservation is mummification, in which tissue is cured with various chemicals and dried. Until recently, the 5000-year-old mummies from Egypt were the oldest known, but 7800-year-old mummies from the northern deserts of Chile have been discovered (Bloch, 1985). In addition to human bodies, the Egyptians preserved a variety of animals, from birds and cats to hippos, bulls, and crocodiles (Hangay and Dingley, 1985*a*). The first attempts at fluid preservation probably were made by the Babylonians, Syrians, and Persians, who immersed their dead in honey to keep them from the air and thus prevent decomposition (Hangay and Dingley, 1985*a*). The Egyptians kept the viscera and brains of human mummies as fluid preparations in a mixture of oils and resins (James, 1957).

The next advancement in fluid preservation came in the mid-17th century. Robert Boyle (1627–1691) discovered that natural history specimens could be preserved in "spirit of wine" (ethyl alcohol). He pickled a linnet and a snake, which were shown before the Royal Society in London (Whitehead, 1970) and given to Gresham College in 1663. The cost of spirit of wine, however, limited the widespread application of his technique (Hangay and Dingley, 1985*b*). Boyle experimented further with preservation in alcohol. In a written communication to the Royal Society of London (Boyle, 1666), he described how he "mingled with the *Spirit of Wine*, a little Spirit of *Sal Armoniack*, made . . . by the help of *Quick-lime*," which he used because he had "never observed it (how strong soever I made it) to coagulate Spirit of Wine." He also found it better to wash the specimens after they had been in solution a while and then move them to a fresh solution for storage.

Another important 17th-century breakthrough was the discovery of a cheap, clear flint glass. Made with a lead oxide, this glass was more transparent than any previously made (Alexander, 1979). With this, it was easier to see the animals preserved in fluid.

Embalming for scientific purposes began to be practiced in Europe in the late 15th century (Hangay and Dingley, 1985*a*). Embalming preserves and disinfects the tissue to stop decomposition due to bacteria and autolysis. Unlike the application of fluids in natural history preservation, embalming fluids usually are pumped mechanically through the circulatory system. The usual method of embalming in Europe at this time was with metallic poisons.

In 1666, Frederick Ruysch preserved the entire body of Vice Admiral Sir William Berkeley with an embalming fluid pumped through the vascular system (Hangay and Dingley, 1985*b*). Ruysch utilized the recent discoveries of William Harvey, whose great work, *An anatomical dissertation on the movement of the heart and blood in animals* had been published in Frankfort in 1628 (Singer, 1957). The fluid Ruysch used was diluted alcohol distilled from fermented barley mash, about 67 percent alcohol by volume (Hangay and Dingley, 1985*b*). The embalmed body of Berkeley kept its fresh appearance until after the death of Ruysch, but did not keep permanently, contrary to Ruysch's expectations.

Despite the cost, the use of spirit of wine as a preservative began to spread. James Petiver (1658–1718) of Aldergate in London printed a sheet of instructions for preserving animals in "Rack, Rum or Brandy" and urged travelers to collect specimens for him (Whitehead, 1970; 1971).

A paper published in London in the middle of the 18th century (Reamur, 1748), describes the techniques then in vogue for treating natural history specimens. The paper describes four methods for "preserving from Corruption *dead birds,* intended to be sent to remote Countries, so that they may arrive there in good Condition." These methods were intended to be applied for use with other vertebrates and insects as well, and included (1) mounting, by removing the skin and either stuffing it or putting it over a mold; (2) putting the specimen "into a Vessel full of Spirit of Wine, or very strong Brandy" (Reamur, 1748), then either keeping it as a fluid specimen or allowing it to dry out (recommended for birds); (3) preparing by "a sort of embalming and even actual embalming, in Countries where the spices are cheap," (Reamur, 1748). For this method, Reamur recommended a powder composed of "resinous Gums, as Aloe, Myrrh, Frankincense, and other Productions of Plants, as Cinnamon, Cloves, Pepper, Ginger, etc. . . . the Result of which will be at least, that the Bird, after being dried, will smell the sweeter, and become as it were a Piece of Perfume." If the spices were prohibitively expensive or unavailable, "you may content yourself with a Salt which is cheap in most countries," or use lime or alum. A fourth preservation method was to "dry them by the Heat of an Oven," (Reamur recommended doing this right after the bread is baked), checking the specimens carefully to make sure they are dried.

The major turning point for collectors in most branches of the natural sciences was the mid-19th-century discovery of formaldehyde. Formaldehyde was first made by the Russian scientist, Alexander Mikhailovich Butlerov, in 1859 (Walker, 1964). While attempting the synthesis of methylene glycol, he noticed the distinctive odor of the formaldehyde gas. It was not until 1868 that August Wilhelm von Hofmann, a chemist at the British Royal Mint, prepared formaldehyde by a process that is the basis for methods of manufacture currently in use. It involves "passing a mixture of methanol vapors and air over a heated platinum spiral" (Walker, 1964). By the 1890s, a botanist, Ferdinand Cohn, found that formaldehyde preserved plants much better than did spirit of wine

(Hangay and Dingley, 1985*b*), and formaldehyde was much less expensive.

Modern fluid preservation of natural history specimens is almost always a two-step process of formaldehyde fixation and alcohol preservation.

CONSERVATION OF FLUID PRESERVED SPECIMENS

Conservation is a new concept for natural history museums, compared to art, history, and anthropology museums. Care of collections always has been of concern in natural history museums, but the techniques used for preparation of specimens are steeped in tradition instead of being based on demonstrated good conservation practices. The rigorous analysis applied by researchers who use specimens in natural history collections has not been applied to the procedures for maintaining the specimens. In some cases, current practices appear to be effective. Other procedures, however, turn valuable specimens into conservation time bombs. Herein lies the paradox of preservation: how can the natural process of organic decay be stopped without altering the nature of the specimen? Romero-Sierra and Webb (1983) stated: "The purpose of any preservation method is to introduce some physical and/or chemical changes to stabilize the tissues, so their structure and composition are preserved in the most life-like state possible. To render desired results, the method used must: block autolitic enzymes, prevent shrinkage, swelling, distortions, dissolution, be bactericidal, light resistant, insoluble, non-flammable, non-toxic, etc. *In reality: the ideal method is a paradox.*" Obviously, because anything done to a scientific specimen, no matter how benign, alters its nature (and thus its value as a scientific specimen), some compromise between preservation and scientific integrity must be reached.

In terms of scientific collections of Recent vertebrates, it is best to look at conservation problems for both field preparation and curation in the museum. In general, conservation measures for fluid collections should be directed toward providing a stable, neutral, fluid preservative for the specimens; reducing ultraviolet (UV) radiation, light intensity, and exposure time; minimizing fluctuations in relative humidity and temperature; and reducing human error.

Although many problems with fluid-preserved specimens are now recognized, we have very little of the data needed to resolve them. Still many other problems go unrecognized year after year. This largely is because the care, or curation of scientific specimens is communicated as an oral tradition and is based almost solely on custom. There are few scientifically tested, rigorous procedures for caring for wet collections.

The very title of "curator" is no longer as descriptive of the position as it once was. Few curators in big museums actually care for the specimens (Colbert, 1958; Laub, 1985). Instead, they direct growth and set policy, leaving curation to an erratically trained staff of technicians. There is no generally recognized body of knowledge that a collection technician is expected to be familiar with. This is compounded by the fact that collections are growing larger (Howie, 1986), while the collections-care staff size is scarcely increasing at all.

In short, natural history museums lack a conservation ethic. The first step in establishing a conservation ethic is to consider the collections in the context of organic deterioration. Objects and specimens, it must be remembered, are series of molecules linked by bonds. When these bonds are broken, changes occur in the physical properties of the molecules. The bonds may be broken by the stress of energy—anything from vibration to light. Following the Third Law of Thermodynamics (entropy), deterioration is dynamic, not static.

In view of this, it should become obvious that conservation of natural history specimens begins with proper documentation of field preparation procedures and continues in the museum in order to document all potential agents of deterioration. The following should always recorded (1) method of killing the specimen; (2) chemicals used in preservation; (3) source of chemicals used (Was formaldehyde purchased locally? Was it diluted with tap water? How did you buffer or neutralize the fixative?); and (4) length of time the specimen was in the fixative and under what conditions (ambient temperature fluctuations; were you traveling by car over bumpy roads?). Documentation in the museum should include (1) condition of the specimen upon arrival; (2) specimen processing procedures (Were the specimens soaked in water? Transfered by stages to standard strength preservative?); and (3) storage conditions.

A conservation ethic must be present at all levels of museum staff, policies, and procedures. Natural history specimen conservation will only become a reality when recognized by all collection users and staff. The collections care staff should be aware that preserved specimens are not stable and that collections are not replaceable. Traditional procedures, materials, and handling techniques should be re-examined in light of what is now known about specimen conservation.

FIXATION OF SPECIMENS

An example of the lack of rigorous thought regarding the selection of fixatives and preservatives traditionally used for fluid specimens can be found in Slevin (1927). Slevin insisted on the use of "pure tin" specimen tags, and specifically rejected the use of paper. He also recommended preserving in alcohol, which he believed "preserves the specimens by drawing water out of their tissues, and by its antiseptic action" (Slevin, 1927). He was completely against the use of formaldehyde, even though he acknowledged ". . . the careless or inexperienced collector who uses it will have less difficulty in preserving his specimens than if he uses alcohol, for specimens placed in formalin never decay. On the other hand, in my judgement, formalin specimens never are as satisfactory as well prepared alcoholic specimens. They usually turn black or a dull leaden gray . . ." (Slevin, 1927). He then noted that his tin tags corroded in formaldehyde, apparently never connecting this chemical reaction to the discoloration.

Slevin's alcohol-fixed specimens at the California Academy of Sciences are still in quite usable condition, and some are more than 80 years old. Most are much more pliable than formalin-fixed material of the same age. I have prepared dry skeletons from toads Slevin collected in 1924, and found the bone to be in good condition. Some of his specimens do not clear and stain very well, however (Simmons, 1987). As discussed below, it may be that Slevin's formalin-free specimens will last several hundred years, whereas his colleagues' formalin-fixed specimens will disintegrate.

Should A Fixative Be Used?

Formaldehyde (HCHO) is a colorless gas. It is used as an aqueous solution of 37 percent formaldehyde gas in water with

12 percent methanol to prevent polymerization (Morrison and Boyd, 1973; Walker, 1964). Formalin, formerly a trade-name, is now used to refer to "full-strength" or aqueous formaldehyde (Pabst, 1987) or this solution further diluted with water, usually 1:9 to make 10 percent formalin (Fink *et al.*, 1979).

Most curators of fluid-preserved collections believe that specimens must be fixed in formaldehyde before they are preserved in some other fluid (Fink *et al.*, 1979; Simmons, 1987). This is a curious belief, in that no fixative, including formaldehyde, has been subjected to rigorous testing to see how long specimens prepared with it will last.

Raikow (1985) stated that *"Fixation* is the treatment of the specimen with chemicals that harden the tissues and prevent their destruction by decay or autocatalytic processes. Unfortunately, the recognition of the need for proper initial fixation is a relatively recent development." He added that "Many specimens from the late nineteenth century or early twentieth century were simply immersed in alcohol after being collected. Tissues in alcohol-fixed specimens tend to be soft and mushy, muscles pull apart easily, and it is difficult to do a clean, clear dissection."

Alberch (1985) defined fixation in terms of possible histological preparation of the specimens. He recommended formalin because it ". . . appears to function as a fixative by forming cross-links between adjacent protein chains, denaturing and thus deactivating them. Autolysis is stopped and proteins are coagulated, preventing breakdown of tissues."

Thus, the concept of fixation that is firmly implanted in the minds of preparators is defined for a specific use (histological preparation) that few of the specimens will actually undergo. The question that curators should be asking is: What, if any, fixative should be used to ensure the conservation of the specimen while still rendering it useful for scientific study? Instead, as Howie (1986) observed, "The very fact that biological material must be 'fixed' or 'preserved' before being worked on in the museum brings about a false sense of security. The techniques of preservation are not designed to maintain specimens eternally. Wet-preserved biological material will probably survive in usable condition for 50 to 70 years, at best. . . ."

It is possible that formaldehyde-fixed specimens kept in alcohol preservative will not last as long as those prepared

without formaldehyde. Alcohol is considered a very poor
fixative, yet, the oldest extant fluid preserved specimens
(which have survived nearly 300 years) were prepared without
formaldehyde, which came into use about 100 years ago
(Hangay and Dingley, 1985b).

Formaldehyde may be turning fluid preparations into time
bombs. Formaldehyde breaks down into formic acid, which
decalcifies bone (Smith, 1947; Taylor, 1977). For this reason,
it must be buffered or neutralized. Even trace amounts of
formaldehyde remaining after specimens have been washed
may cause acidification of the alcohol preservative, which will
then decalcify bone (Dingerkus, 1982). Because formaldehyde oxidizes easily, it should only be diluted with distilled
water. This rarely happens in practice.

The problem of acidification of formaldehyde solutions has
been recognized for a long time, but little attention is paid to
it in natural history collections. Smith (1947) pointed out
that like all aldehydes, formaldehyde absorbs oxygen readily,
even from the air, thus quickly reaching a pH of 3.5 or even 3.
He cautioned that calcified and ossified structures are decal-
cified in the acidic solution. A neutral or slightly alkaline formaldehyde solution is desirable. Smith also pointed out the
undesirability of borax (the traditional buffer) for long-term
use with fluid specimens.

Buffering or Neutralizing Formaldehyde

There is some evidence from the tanning industry that formalin may fix better in an acidic environment, but due to its
tendency to break down to formic acid, it should be buffered
or neutralized. Even this process is not without problems—
formaldehyde can be made too alkaline, and the alkaline environment would then react with the proteinaceous material.

The traditional buffer for formaldehyde, borax, works initially, but has been shown to actually assist in acidification of
the preserving solution over time (Fink et al., 1979; Taylor,
1977). Recommendations to use more stable buffers are
generally not followed. Taylor (1977) suggested calcium carbonate instead of borax. Alberch (1985) recommended buf-
fering with sodium acid phosphate ($NaH_2PO_4H_2O$) and
anhydrous disodium phosphate (Na_2HPO_4) for better histological fixation. Quay (1974) recommended "a precise
mixture of slightly acidic and basic salts" instead of calcium

carbonate, magnesium carbonate, or marble chips, as "these can give rise to hard crystalline deposits in muscle and other tissues . . ." making them inadequate for histological studies.

Once the specimens are formaldehyde fixed, many authors, such as Jones and Owen (1987), recommend washing specimens to get formaldehyde out, as it may otherwise cause the alcohol solution to become acidic (Dingerkus, 1982). To avoid having enzymatic activity reoccur (Taylor, 1981), it is better to wash the specimens through a series of alcohol dilutions up to the desired storage strength.

Another caution regarding the use of formaldehyde is the potential health hazard it presents to the museum staff. Next to the pesticides used in bird and mammal collections, formaldehyde is probably the most dangerous chemical in use in natural history museums today. In addition to causing irritation of the nasal passages and contact dermatitis, it is a carcinogen (NIOSH/OSHA, 1980; Pabst, 1987; Perera and Petito, 1982). It will penetrate gloves of natural rubber, polyvinyl chloride, or polyethylene (Raloff, 1985). Formaldehyde resistant gloves are now available in thicknesses of 10–22 mil. Soft contact lenses may absorb the vapors, and vapors may become trapped behind hard contact lenses (Cohen *et al.*, 1979). Work areas where formaldehyde is used should be well ventilated or vapor masks should be worn (Hearne and Catania, 1983).

Before the use of formaldehyde as a fixative is abandoned, however, research must be done to determine (1) if fixation is necessary, and (2) if so, is there a better fixative than formaldehyde that should be used.

PRESERVATION OF SPECIMENS

Fluid

What is the best fluid for preserving fixed or unfixed specimens? As mentioned, "spirit of wine" (ethyl alcohol) was first recommended in the mid-1600s. It is still in use. A preservative is supposed to prevent microbial growth, as well as chemical or physical changes in the specimen that will interfere with later study of it (Quay, 1974). Alcohol is a good biocide, but it also dehydrates tissue. Various authors have recommended strengths ranging from 60 percent to 75 percent ethyl alcohol (ETOH). There are no data to indicate which concentration is best, other than it is known that a 70 percent ethyl alcohol solution will penetrate bacterial cells.

The second most common preservative is isopropyl alcohol, usually in strengths of 35 percent to 50 percent. This is the only preservative that has been tested in a controlled situation (Lai, 1963). This study showed that isopropyl alcohol causes significant changes in proportional measurements of fish specimens, compared to ethyl alcohol.

Ellis (1987) reviewed reports on biomass loss in vertebrates and invertebrates in several studies. Biomass loss varied greatly with different methods of fixation and preservation and the resultant leaching of materials from the specimens.

Tucker and Chester (1984) conducted experiments on twelve different fixative/preservative solutions for larval fish over a seven year period. Based on experimental evidence, they recommended four percent formaldehyde in distilled water, buffered with one percent sodium acetate, for both fixation and preservation of larval fish.

Alcohol can become acidic over time and decalcify specimens, particularly if it was not diluted with distilled water (Dingerkus, 1982). Trace amounts of formaldehyde in fixed specimens stored in alcohol preservative can also speed acidification of the alcohol. Crowding of specimens in a container, or the addition of freshly fixed specimens to a container of alcohol will dilute the alcohol concentration (Taylor, 1981), sometimes to dangerously low levels.

Other Materials in Contact with Preservatives

Materials for specimen tags are another consideration. Jones and Owen (1987) caution that aluminum bat bands will deteriorate rapidly in formalin and alcohol, and Hawks and Williams (1986) found that dissimilar metals in contact in the presence of an electrolyte (such as formalin or alcohol) will corrode. The additives used in plastic labels (ultraviolet light absorbers, anti-oxidants, plasticizers, etc.) may be volatile and have the potential to cause harm to specimens (Hawks and Williams, 1986). Hawks and Williams (1986) concluded that paper is preferred for specimen tags and labels, but it has its own problems. It can be inherently weak, may have been manufactured with an acidic sizing, may contain lignin (an organic acid), may contain bleaches or dyes, and it is susceptible to fluctuations in temperature and relative humidity.

They recommend 100 percent cotton stock, neutral to mildly acidic (pH of 6.5 to 7.0). They caution against using papers with an alkaline buffer, because the alkaline materials may themselves degrade the protein of the specimen.

Williams and Hawks (1986) found many kinds of ink in use in museums not to be formaldehyde or alcohol resistant. For example, the most commonly used ink for vertebrate collections for over 20 years has been Higgins Eternal Black, but it did not rate as an acceptable ink for permanant records in their tests. They provide a list of acceptable inks.

When glass vials are used in fluid preservative, they should be plugged with cotton. Corks give off tannic acid (Levi, 1966), plastic caps become loose and float free from the vials, and rubber or similar material either becomes hard or dissolves in the preservative, contaminating the specimens.

Alcohol and formaldehyde should be diluted with distilled water. The use of tap water, which is very common, usually results in some precipitate forming in the solution. Depending on the contaminants in the tap water, the precipitate could discolor or damage specimens.

Loss of Color

Fluid preservation in ethyl alcohol will result in the loss of certain colors in the specimens. For example, some pigments in bird feathers, such as lipochromes (which produce yellow, orange, and red colors in feathers), are alcohol-soluble carotenoids (Welty, 1979). Several formulas for preservatives with better color retention than formalin have been proposed (Guerra, 1976; Waller and Eschmeyer, 1965; White and Peters, 1969; and Windsor, 1971), but none have been subjected to long-term tests.

Cleared and Stained Specimens

Cleared and stained specimens are kept in full-strength glycerine. A few crystals of thymol ($C_{10}H_{14}O$) must be added to prevent fungal growth. It is not known how long cleared and stained specimens will endure, but the specimens are very fragile after undergoing this process and appear to disintegrate fairly rapidly.

OTHER CONSERVATION CONSIDERATIONS

Fluid preservation of specimens eliminates some problems faced by other types of collections, such as pest infestation, but it has unique difficulties of its own that are associated with several environmental factors. The importance of a stable environment for the collection cannot be stressed too much. With each fluctuation of temperature, humidity, or both, the specimens and specimen containers in the collection will make corresponding movements.

Light

According to Lull and Merk (1982), "Light is a silent destroyer." Light is a form of energy, and thus causes the breakdown of organic materials. Both the wavelength and the amount of exposure time must be considered; light damage is cumulative. In particular, ultraviolet light (UV) should be reduced, because "ultraviolet radiation causes fading, discolouration and structural damage such as embrittlement and loss of tear-strength," (Harris, 1968) in organic materials. A rise in temperature of 10°F may double the rate of UV activity (Stolow, 1966*b*).

The most damaging ultraviolet light is in the wavelengths of 300 nm (nanometers) to 400 nm (Lull and Merk, 1982). Tungsten filament bulbs emit only a negligable amount of UV light. They do produce heat, but usually not enough to be a problem. Ordinary glass is inadequate to filter out UV light (Harris, 1968), as it allows light of 310 nm to 400 nm to pass through (Macleod, 1975). Specimens and labels in jars of fluid should never be exposed to sunlight, even sunlight coming through window glass.

Many collections of fluid preserved specimens are exposed to damaging levels of ultraviolet radiation from fluorescent tubes, but UV light can be controlled with filters and by reducing exposure time. Filters must be able to absorb radiation of wavelengths less than 400 nanometers, and are usually made of acrylic and polyvinylchloride (PVC) plastics, sometimes laminated glass (Harris, 1968). In addition, fluorescent tubes have an abnormally high light output during the first 100 hours or so of operation, so a burn-in period is recommended outside the collection storage area during their first 100 hours of operation (Lull and Merk, 1982).

Light levels are usually kept too high in museums. Lights in collection rooms should be controlled by aisle or bank, so that only those lights that are needed are turned on. It is also important to reduce light levels around field notes, rare books, catalogs, and so forth.

Evaporation and Dehydration

In terms of fluid preserved collections, evaporation means that the vapor pressure "pushes" the fluid out of a jar. Most evaporation occurs around the seal between the jar and the lid. Consider a collection of 50,000 jars with an average lid circumference of 30 cm, the size of a quart jar. This means the collection would have a 15-kilometer long line of evaporation potential.

Different materials have different thermal expansion characteristics, which is why metal or plastic lids on glass jars will "back off" or unscrew themselves if subjected to even a small variation in temperature. Metal lids eventually will rust and allow evaporation to take place. Most rubber and rubberlike gaskets age rapidly in alcohol, become dry, inflexible, and crack (Fink *et al.*, 1979). Due to a difference in vapor pressure, alcohol evaporates faster than water. A jar with a bad seal may appear to have enough liquid in it, but actually be far below 70 percent ETOH, endangering the specimens. Preservative concentrations must be checked periodically with a hygrometer or, better yet, a density meter, which is much more accurate.

Alcohol-preserved specimens are hygroscopic. While being examined, they should be kept in a tray of the appropriate preservative. They should never be put in water. Water absorption will damage tissues, and the specimens will dilute the preservative when returned to the jar (Simmons, 1987; Taylor, 1981). A dehydrated specimen cannot usually be rehydrated without damaging the specimen in the process.

Temperature

Temperature is a measure of the rate of movement of atoms and molecules, or their kinetic energy. As temperature increases, a greater proportion of any given substance will be in its most mobile phase (that is, for alcohol, a gas instead of a liquid). For a fluid-reserved specimen in a jar, this means in theory that the body oils migrate out of the specimen faster, the acidic preservative diffuses into the bone and thus the calcium diffuses out faster, and the greater number of molecules in the air space above the fluid level means a greater loss of preservative from the jar. Thus, a cooler temperature presumably slows deterioration processes.

Collection room temperature should be maintained at 65°F ± 5° (Cato, 1986; Fink *et al.*, 1979; Stolow, 1966*a*) as mold growth occurs at 70°F. Maintenance of a constant temperature is a primary consideration. Changes in temperature and relative humidity are stressful to material because they cause expansion and contraction. Cooler temperatures are better for organic materials than higher ones, as decay reactions occur faster at higher temperatures. Heat also contributes to color loss and general degredation of preserved specimens.

Conservation of Specimen Documentation Records

Considering the importance of field notes to natural history collecting, it is curious that so little attention has been given to documenting specimen preparation (Cannell *et al.*, 1988). The preservation history of the specimens should be documented, including anything that is done to them once they are in the collection (Cannell *et al.*, 1988; Jones and Owen, 1987). The advent of electronic data management systems in museums has made it easier to keep track of this kind of information.

Few workers in natural history museums are aware of how impermanent their library resources and written records are. Books, journals, field notes, specimen labels, tape recordings, photographs, and electronically stored data are all susceptible to damage by heat and moisture (Banks, 1978; Browning, 1970). All of these are inherently unstable, because they are composed (at least in part) of organic compounds. Only specially made paper will last more than about 50 years, because the acid contained in paper catalyzes hydrolytic degradation of polymeric cellulose molecules (Shahani and Wilson, 1987). The process of deterioration can be slowed by keeping cellulosic materials under conditions of cool temperature and low humidity (Banks, 1978).

Recorded tapes and electronically stored data (in any form) are even less stable than paper, as they can be destroyed by static electricity, exposure to a magnetic field, or improper placement in storage (McWilliams, 1979), in addition to heat and moisture.

Natural Disasters

Plans should be made for dealing with natural disasters that might befall the collection such as earthquakes and fires. See Jones (1986) and Waddington and Fenn (1986) for information on safety and emergency planning for museums. In particular, contingency plans should be made for loss of preserving fluid, loss of collection documentation, and loss of storage containers.

Use of Collections

For conservators of art, history, or anthropology collections, the consumptive use of specimens is prohibited by their code of ethics (AIC, 1988). This is a conservation issue because consumptive use alters the physical integrity of the object (or specimen). In natural history collections consumptive uses of some specimens are fairly common. Traditionally, the decision to allow such uses of material has been made by the curator of the collection. It generally is based on the number of specimens of a particular species in the collection, the availability of similar specimens in other collections, the replaceability of the specimens, and the value of the information to be gained from the proposed research.

The nature of the research done with natural history specimens is very different from that done with art, anthropological, or historical objects. A suggestion to prohibit the consumptive use of natural history specimens in research would be met with derision in the scientific community. A more realistic and useful goal is to develop guidelines for better evaluation of consumptive uses of specimens than are sometimes used at present.

Future Use of Collections

The application of new systematic techniques using proteins or DNA, for example, (Barrowclough, 1985) will not be possible for some fluid-preserved specimens due to the lack of documentation concerning killing, fixation, preservation, and conditions of storage. New fixation and preservation techniques will probably be required for some new systematic techniques (Ouellet, 1985). These should be developed with the goal of specimen conservation in mind.

The Future of Conservation
in Natural History Museums

Howie (1986) noted that "Little research has taken place on the development of new techniques and materials that might be used in the preservation of natural history specimens. Much appears to have been borrowed from the field of pathology, in which short-term preservation is the general requirement. . . . Many techniques are so steeped in tradition that today's practitioners often have no understanding of their basis."

Even though fluid preserved natural history specimens exist principally for scientific research, it is a sad fact that most of the approach to fixation and preservation is not scientific. It has been suggested, for example, that gluteraldehyde might be a better fixative than formaldehyde. Even though this has not yet been tested, some museums are already changing to gluteraldehyde.

What is needed for fluid-preserved natural history collections is:

1. Research in the following areas.
   a. Fixatives—is a fixative necessary, and if so, which one will allow the maximum useful lifespan of the specimen? Which if any fixatives and buffers currently in use are damaging to specimens and should no longer be used?
   b. Preservatives—which is best for the long-term survival and usability of the specimen?
   c. What is the best combination of paper and ink for fluid collections, especially for computer generated labels?
2. A conservation code of ethics for natural history specimens should be established. The code of ethics currently accepted for conservators (AIC, 1988) is, for the most part, readily adaptable to workers in natural history collections. Briefly, it requires competence on the part of conservators, good facilities, use of a single standard for all objects (specimens); suitability of treatments undertaken; limitations of aesthetics on reintegration, the continued

self-education of conservators, and the oversight of auxiliary personnel working with the collections. There are potential conflicts, for natural history collections, with the principle of respect for the integrity of the object (specimen), and the principle of reversibility, due to the nature of the research undertaken with natural history collections.

3. There is a need for support for conservation efforts from curators and administrators.
4. Materials (and the names of the manufacturers) that are used in the field and laboratory should be documented. Write down how you actually prepared the specimens, not a general description of how you should have.
5. There is a need to break with "traditional" methods of fixation and preparation.
6. Encourage development of conservation societies, journals, and research.

Conservation has a low profile in natural history museums. It is highly unlikely that those writing the grants for collections support (almost always curators who are running their own research programs) will begin to request funds for conservation until they can be made aware of the problems facing the collections. Funds for the support of systematic collections are limited. New funding sources for specimen conservation need to be developed.

CONCLUSION

Natural history collections have great value, but there is little money to care properly for them. Practically no money is available for conservation of natural history specimens, especially compared to art collections (Howie, 1986). Despite the value of the collections in their care, most curators are unaware of the serious conservation problems they face. Collections have grown at a phenomenal rate during the last 50 years. It is estimated that worldwide, there are some 1.5 billion specimens in collections, and these collections are growing by about 50 million specimens a year, yet "a significant proportion of this vast resource is in poor condition or actively deteriorating, even in eminent national collections" (Howie, 1986).

Acknowledgments

I wish to thank the participants in the 1987 Collections Care Pilot Training Program (CCPTP), especially Lynn Barkley, Mary Lou Florian, Catharine Hawks, Kimball Garrett, Sally Shelton, Arnold Suzumoto, and Robert Waller. The provocative ideas and stimulating discussions of the CCPTP provided the basis for this paper, though I take full responsibility for the interpretations presented here. I would also like to thank Cathy Dwigans for her critical comments on this manuscript and her support and encouragement of my interest in natural history conservation issues.

Literature Cited

AIC. 1988. Code of ethics and standards of practice. Pp. 20-30 *in* The American Institute for Conservation of Historic and Artistic Work. AIC, Washington, D.C. v+1-184 pp.

Alberch, P. 1985. Museum collections and the evolutionary study of growth and development. Pp. 29-41, *in* Museum collections: their roles and future in biological research (E. H. Miller, ed.). British Columbia Provincial Museum Occasional Papers Series, 25:x +1-219.

Alexander, E. P. 1979. Museums in motion. Amer. Assoc. State Local Hist., Nashville, xii + 308 pp.

Banks, P. N. 1978. Preservation of library materials. Pp. 180-222 *in* Kent, A., H. Lancour and J. E. Daily (eds). Encyclopaedia of library and information science, Vol. 23. Maracel Dekker, Inc., New York. iv+1-512 pp.

Barrowclough, G. F. 1985. Museum collections and molecular systematics. Pp. 43-54, *in* Museum collections: their roles and future in biological research (E. H. Miller, ed.). British Columbia Provincial Museum Occasional Papers Series, 25:x + 1-219.

Bedini, S. A. 1965. The evolution of science museums. Technology and Culture, 6(1):1-29.

Bloch, N. 1985. In the battle of the embalmers, nature triumphs. Earthwatch News, 7(1):2.

Boyle, R. 1666. A way of preserving birds taken out of the egge, and other small faetus's. Philos. Trans. Royal Soc. London, 12:199-201.

Browning, B. L. 1970. The nature of paper. Pp. 18-38 *in* Winger, H. W. and R. D. Smith (eds). Deterioration and preservation of library materials. Univ. Chicago Press, Chicago, 200 pp.

Cannell, P. F., M. R. Bakst and C. S. Asa. 1988. Suggestions regarding alcoholic bird collections. Condor, 90:500-503.

Cato, P. S. 1986. Guidelines for managing bird collections. Museology, 7:1-78.

Cohen, K. S., J. L. Ennis, I. Arons, A. J. Krubsack, R. D. Rowe, G. C. Lowry. 1979. Caution to contact lens users [Letters]. Chemical and Engineering News, 57(47):4; 84.

Colbert, E. H. 1958. On being a curator. Curator, 1:7-12.

Dingerkus, G. 1982. Preliminary observations on acidification of alcohol in museum specimen jars. Curation Newsletter, 5:1-3.

Ellis, D. V. 1987. Biomass loss in wet-preserved reference collections. Collection Forum, 3(1&2):6-8.

Fink, W. L., K. E. Hartel, W. G. Saul, E. M. Koon, and E. O. Wiley. 1979. A report on current supplies and practices used in curation of ichthyological collections. ASIH *ad hoc* subcommittee report, 63 pp.

Guerra, L. A. 1976. Color preservation in salamanders. Herpetol. Rev., 7(4):170-171.

Hangay, G. and M. Dingley. 1985*a*. Biological museum methods. Volume 1. Vertebrates. Academic Press, 379 pp.

———. 1985*b*. Biological museum methods. Volume 2. Plants, invertebrates and techniques. Academic Press, 323 pp.

Harris, J. B. 1968. Practical aspects of lighting as related to conservation. Pp. 133-138, *in* Contributions to the London Conference on Museum Climatology. International Institute for Conservation of Historic and Artistic Works, London. 1-296 pp

Hawks, C. A. and S. L. Williams. 1986. Care of specimen labels in vertebrate research collections. Pp. 105-108, *in* Proceedings of the 1985 Workshop on Care and Maintenance of Natural History Collections (Waddington, J. and D. M. Rudkin, eds.). Life Sciences Misc. Publ., Royal Ontario Museum. v+121 pp

Hearne, M. and D. Catania. 1983. Formaldehyde vapor masks. Curation Newsletter, 6:7.

Howie, F. M. P. 1986. Conserving natural history collections: some present problems and strategies for the future. Pp. 1-6, *in* Proceedings of the 1985 Workshop on Care and Maintenance of Natural History Collections (Waddington, J. and D. M. Rudkin, eds.). Life Sciences Misc. Publ., Royal Ontario Museum. v+121 pp.

James E. O. 1957. Prehistoric religion. Barnes and Noble, Inc., New York, 300 pp.

Jones, B. G. (ed). 1986. Protecting historic architecture and museum collections from natural disasters. Butterworths, Boston, xi + 560 pp.

Jones, E. M. and R. D. Owen. 1987. Fluid preservation of specimens. Pp. 51-63, *in* Mammal collection management (H. H. Genoways, C. Jones and O. L. Rossolimo, eds.). Texas Tech Univ. Press, Lubbock, iv + 219 pp.

Lai, Y. 1963. Effects of several preservatives on proportional measurements of the fat-headed minnow, *Pimephales promelas*. Unpubl. M.A. thesis, Univ. Kansas, Lawrence, 40 pp.

Laub, R. S. 1985. The natural history curator: a personal view. Curator, 28:47-55.

Levi, H. W. 1966. The care of alcoholic collections of small invertebrates. Syst. Zool., 15:183-188.

Lull, W. P. and L. E. Merk. 1982. Lighting for storage of museum collections. Technology and Conservation, 7(2):20-25.

Macleod, K. J. 1975. Museum lighting. Canadian Conservation Institute Technical Bulletin, 2:1-13.

McWilliams, J. 1979. The preservation and restoration of sound recordings. Amer. Assoc. State Local Hist., Nashville, 138 pp.

Morrison, R. T. and R. N. Boyd. 1973. Organic chemistry. Third edition. Allyn and Bacon, Inc., Boston, ix + 1258 pp.

NIOSH/OSHA. 1980. Formaldehyde: evidence of carcinogenicity. NIOSH/OSHA Current Intelligence Bulletin, 34:1-15.

Ouellet, H. 1985. Museum collections: perspectives. Pp. 215-219, *in* Museum collections: their roles and future in biological research (E. H. Miller, ed.). British Columbia Provincial Museum Occasional Papers Series, 25:x + 1-219.

Pabst, R. 1987. Exposure to formaldehyde in anatomy: an occupational health hazard? The Anatomical Record, 219:109-112.

Perera, F. and C. Petito. 1982. Formaldehyde: a question of cancer policy? Science, 216:1285-1291.

Quay, W. B. 1974. Bird and mammal specimens in fluid— objectives and methods. Curator, 17:91-104.

Raikow, R. J. 1985. Museum collections, comparative anatomy and the study of phylogeny. Pp. 113-121, in Museum collections: their roles and future in biological research (E.H. Miller, ed.). British Columbia Provincial Museum Occasional Papers Series, 25:x + 1-219.

Raloff, J. 1985. Unwanted formaldehyde 'breakthrough.' Science News, 127:297.

Reaumur, R. A. F. 1748. Diverse means for preserving from corruption dead birds, intended to be sent to remote countries, so that they may arrive there in good condition. Some of the same means may be employed for preserving quadrupeds, reptiles, fishes and insects. Philos. Trans. Royal Soc. London, 45:304-320.

Romero-Sierra, C. and J. C. Webb. 1983. The potentials of diatirology. Pp. 21-28, in Proceedings of the 1981 Workshop on Care and Maintenance of Natural History Collections (D. J. Faber, ed.). Syllogeus, 44:21-28.

Shahani, C. J., and W. K. Wilson. 1987. Preservation of libraries and archives. American Scientist, 75:240-251.

Simmons, J. E. 1987. Herpetological collecting and collections management. Society for the Study of Reptiles and Amphibians Herpetological Circular 16, 70 pp.

Singer, C. 1957. A short history of anatomy and physiology from the Greeks to Harvey. Dover Publications, New York, xii + 209 pp.

Slevin, J. R. 1927. The making of a scientific collection of reptiles and amphibians. Proc. Cal. Acad. Sci., 16:231-259.

Smith, J. L. B. 1947. A neutral solution of formaldehyde for biological purposes. Trans. Royal Soc. South Africa, 31:279-282.

Stolow, N. 1966a. The action of environment on museum objects. Part I: humidity, temperature, atmospheric pollution. Curator, 9:175-185.

————. 1966b. The action of environment on museum objects. Part II: light. Curator, 9:298-306.

Taylor, W. R. 1977. Observations on specimen fixation. Proc. Biol. Soc. Washington, 90:753-763.

Taylor, W. R. 1981. Preservative practices: water in tissues, specimen volume, and alcohol concentration. Curation Newsletter, 2:1-3.

Tucker, J. W. and A. J. Chester. 1984. Effects of salinity, formalin concentration and buffer on quality of preservation of southern flounder (Paralichthys lethostigma) larvae. Copeia, 1984:981-988.

Waddington, J. and J. Fenn. 1986. Health and safety in natural history museums: an annotated reading list. Pp. 117-121 in Proceedings of the 1985 Workshop on Care and Maintenance of Natural History Collections (Waddington, J. and D. M. Rudkin, eds.). Life Sciences Misc. Publ., Royal Ontario Museum. v+121 pp

Walker, J. F. 1964. Formaldehyde. Third ed. American Chemical Society Monograph Series, Reinhold Publishing Company, New York, xi + 397 pp.

Waller, R. A. and W. N. Eschmeyer. 1965. A method for preserving color in biological specimens. Bioscience, 15:361.

Welty, J. C. 1979. The life of birds. Second ed. Saunders College Publishing, Philadelphia, xv + 623 pp.

White, D. A. and E. J. Peters. 1969. A method of preserving color in aquatic vertebrates and invertebrates. Turtox News, 47:296-297.

Whitehead, P. J. P. 1970. Museums in the history of zoology. Museums Journal, 70:50-57.

———. 1971. Museums in the history of zoology. Museums Journal, 70:155-160.

Williams, S. L. and C. A. Hawks. 1986. Inks for documentation in vertebrate research collections. Curator, 29:93-108.

Windsor, D. A. 1971. Ammonium sulfate as a preservative which does not remove color from frogs. Copeia, 1971:356-357.

Natural History Museums: Directions for Growth
Paisley S. Cato and Clyde Jones, editors
Texas Tech University Press, Lubbock, 1991, iv+252 pp.

POLICIES CONCERNING THE USE AND MANAGEMENT OF ANCILLARY PREPARATIONS IN VERTEBRATE SYSTEMATICS COLLECTIONS

Paisley S. Cato and David J. Schmidly

Abstract.—Ancillary preparations is a catch-all phrase which, for vertebrate collections, commonly refers to all preparations other than traditional ones (study skins and skulls, skeletons, and fluid-preserved specimens). With the increased emphasis on molecular studies in systematics as well as the demand for maximizing data from each specimen, both the quantity and importance of ancillary preparations have grown dramatically. Some of the preparations can be managed according to existing collection management procedures; others, such as frozen preparations, pose unique problems that must be addressed. The practical and ethical concerns surrounding these materials demonstrate the need for effective management policies and procedures to standardize practices and ensure equitable access to the material. The Mammal Division of the Texas Cooperative Wildlife Collection (TCWC) has developed a working policy to govern the use and management of ancillary preparations, based on accepted museum collection management theory. The policy addresses the formation of an advisory committee and policies for acquisition, record keeping, specimen use and deaccessions. Specimen use is categorized into three types (loans, gifts, and exchanges) and provisions are made to restrict the use of certain preparations.

Many natural history collections now face the prospect of managing new types of specimens for which professional standards of use and management have not been fully developed. Many of these specimens have resulted from the application of molecular studies and other biotechnology to systematics during the past 30 years. However, it has only been during the past decade that the need to develop and manage permanent repositories for these materials has gained broader support (Dessauer and Hafner, 1984).

These relatively new specimens include a variety of frozen and nonfrozen preparations, and often are identified as "nontraditional preparations" or "ancillary preparations" (Table 1). They frequently accompany and complement the traditional vertebrate preparations which include skins, skulls, skeletons, and fluid-preserved materials. Ancillary preparations present a number of practical (or procedural) and ethical concerns that must be addressed by professional museum and scientific societies to ensure adequate care and equitable use of the samples.

TABLE 1.—*Summary listing of ancillary preparations.*

NON-FROZEN PREPARATIONS

Parasites
Photographs, slides and negatives
Karyotype slides, histological slides
Special fluid-preserved samples
    (phalli, hearts, brains, cleared and stained)
Stomach contents, feces, owl pellets
Field notes
Sound recordings
Plaster track casts
Sectioned teeth
Hair samples

FROZEN PREPARATIONS

Collected directly from specimen
    Tissues
        (heart, liver, muscle, kidney, eyes, other)
    Blood components
        (whole, serum, antisera, plasma)
    Whole animals
    Semen
    Venom

Result of laboratory processing
    Isolated nucleic acids
    Cloned DNA probes
    Isolated proteins
    Tissue homogenate
    Cell suspensions

Procedural and Practical Concerns

Although practical concerns generally are considered a procedural or technical matter, they influence the policies that guide the development of ancillary preparation collections. For ancillary collections, the nature of the specimen preparation determines to a great extent the ease with which material can be managed. Standard management and conservation procedures already exist for some ancillary preparations such as slides, photographs (Haas, 1983; Rempel, 1983; Ritzenthaler *et al.*, 1984), sound recordings (Hardy, 1984; McWilliams, 1979), and parasites (Pritchard and Kruse, 1982). The care of most other non-frozen preparations can be based on professional standards for similar types of material. For example, the management and conservation of field notes

there is the continual cost of providing electricity, or replenishing liquid nitrogen as it evaporates from tanks. Finally, there is the expense of a security system, which should include an alarm for alerting personnel when power is disrupted or the temperature changes ($600 to $1500 for initial cost), plus a backup system, such as another freezer or a generator to provide emergency power ($1000 to $2500 for a manually started unit).

Installation and removal of materials must be quick and well-planned to avoid warming up the freezer. The interior of the unit must be efficiently organized to permit both maximum storage capacity and ease of installation and retrieval.

A final area of operational concern that must not be overlooked is the additional personnel time and expertise required to manage these collections. Frozen tissue collections are not as "forgiving" as traditional collections from the management perspective and require closer surveillance. Samples can not be left out on work tables while inventories and computerized data bases are updated. Inventories change more rapidly than for traditional materials and must be managed on a daily basis. Equipment must be routinely monitored. The personnel commitment required to manage a growing frozen tissue collection should be considered as time added to existing personnel time and not merely absorbed by existing staff. This means increased operational costs, adding to the total expense of operating frozen tissue collections.

ETHICAL CONCERNS

The ethical concerns affecting the collection, management, and preservation of ancillary preparations are generally similar to those of traditional vertebrate collections (Cato, 1986; Simmons, 1987; Williams et al., 1977). However, some of the issues are exacerbated for frozen preparations, partly because of the nature of the material and how it is used. Two primary areas of concern are equitable access to material and the conflict between consumption of samples for current research and conservation of samples for future reference.

Tissue and blood samples taken from many vertebrate species normally are available in very small quantities, and, therefore, can be used for a finite number of studies. Decisions concerning who can use the material and for what purposes become critical to satisfy the goal of obtaining the best information from a limited sample. In addition, there is a need to preserve samples of genetic diversity for future studies that might involve a different scientific perspective and new technologies. Although the preservation of samples of genetic diversity is cited as particularly important for rare or endangered species, the philosophy holds true for more common species as well. Technologically, it is possible to preserve samples indefinitely if they are properly collected and stored (Dessauer and Menzies, 1984); practically, individuals responsible for collection care must balance the needs of current research with those of the future.

Who should determine access to samples and on what basis? Is it acceptable to arbitrarily decide that only one-half of the tissue for a specimen should be used for immediate research? Should the decision concerning quantity have more stringent restrictions? What restrictions should apply and for which species and specimens? These are all questions that have yet to be fully addressed.

POLICIES

The need for policies to guide the development, use, and management of ancillary preparations becomes more evident in light of the procedural and ethical concerns discussed above. The goal of an institutional policy should be to establish guidelines that will balance the concerns for use and conservation. The policy should establish standards that will provide continuity when personnel change. The process of developing a policy can be as valuable as the final product because it forces collection personnel to question why and how decisions are made.

In 1987, as part of a process to develop the ancillary preparation collection, the staff of the TCWC mammal division developed an institutional policy for the use and management of ancillary preparations (Appendix). The concepts addressed in the policy reflect the current theory for museum collection management policies as discussed in Malaro (1979, 1985), Porter (1986), and Cato (1986) as well as concerns expressed in Lee *et al.* (1982) and Dessauer and Hafner (1984). The initial draft of the policy was reviewed by several independent researchers before being adopted as a working policy. The TCWC policy is a working document and will be revised as the collection develops and professional standards evolve.

Although the policy focuses on the situation at Texas A&M University, it has been included in this paper to stimulate discussion among professionals working with these collections. Institutional policies should reflect professional standards while serving the operational structure of the individual situation. However, at this time, professional standards have not been widely adopted.

Among the points included in the TCWC policy which should be particularly addressed are: (1) the formation of an advisory committee to monitor the use and maintenance of materials, (2) the definition of specimen use, and (3) the policy concerning restrictions on use. The primary goal of the advisory committee is to ensure equitable access to materials. It is important to consider a balance between the needs of the individual who collected the material under the auspices of a specific research project, the needs of other researchers, and the need for conserving material for future research. Individuals responsible for managing collections are bound ethically to manage the contents as a trust for the public (Force, 1975; American Association of Museums, 1978; Malaro, 1985). A collection, such as the TCWC, which serves as a repository for many researchers, must develop a mechanism to make informed decisions concerning the use of collection resources. The advisory committee provides a balance of viewpoints. The committee is comprised of two members from the collection staff, two independent researchers on the Texas A&M University campus, and the collector or researcher whose material is under consideration. Obviously, this last position is a revolving, temporary position, dependent on the material being reviewed.

The advisory committee makes recommendations concerning acquisitions, based on several criteria: rarity of the taxon, its geographic origin, research programs, current collection holdings, and the presence of traditional vouchers. It also deals with requests to use samples, and the implementation of restrictions on certain materials. The curator of mammals makes the final decision in all of these areas, taking into consideration the recommendations of the committee.

The TCWC policy categorizes specimen use into one of three types: loans, gifts, and exchanges. As Baker and Hafner

(1984) point out, much of the material sent out to researchers is consumed during analysis, and nothing is left to return to the home institution. The distinction between loans, gifts, and exchanges should be obvious, but what is less apparent is the record keeping and documentation that should accompany each type of transaction. If the collection inventory is to be accurate, these uses must be uniquely identified and recorded as such. Thus, it is critical to make a distinction among types of use at the policy level so that emphasis is placed on procedural standards to reflect the transactions accurately.

It is commonly agreed that restrictions should be placed on the use of some contents of a frozen tissue collection (Baker and Hafner, 1984; Dessauer et al., 1988). The difficulty arises in trying to decide which species and which specimens, how much, and for how long. Certainly, there are no absolute rules that can be used to establish restrictions on material. The TCWC policy uses the advisory committee to assist in this process by making recommendations concerning restrictions to the curator of mammals. These restrictions are dealt with on a case-by-case basis, usually when the material is deposited into the collection. The restrictions can be reviewed at any point and revised if necessary.

Among the professional societies that are addressing the problems of frozen materials are the American Society of Mammalogists (ASM), the Society for the Preservation of Natural History Collections (SPNHC), and the Association of Systematics Collections (ASC). In 1989, the Systematics Collections Committee of ASM began a survey to identify the types of mammalian materials held as ancillary preparations and the procedures followed to manage them. This information will form the basis for a future workshop and publication. ASC and SPNHC have focused on these questions at workshops and through individual presentations at recent annual meetings.

SUMMARY

The importance of developing comprehensive, flexible policies for the use and management of ancillary preparations can not be over stressed. The potential for problems to arise is great and the development of policies in advance of the problem will help alleviate many difficult situations. Dessauer et al.

(1988) proposed some considerations for the development of national repositories, but the issues need to be addressed at the institutional level before an effective national program can be instituted. Individual institutional policies, such as the one presented with this paper, can serve as a stimulus for discussion and lead to the development of discipline-wide standards.

Acknowledgments

Funding for the development of the Texas Cooperative Wildlife Collection Ancillary Preparation Collection has been provided by the National Science Foundation (BSR-8709724) and the Texas A&M University System.

Literature Cited

American Association of Museums. 1978. Museum ethics. A report to the American Association of Museums by its Committee on Ethics. Washington, D.C., 31 pp.

Baker, R. J., and M. S. Hafner. 1984. Curatorial problems unique to frozen tissue collections. Pp. 35-40 in Collections of frozen tissues: value, management, and directory of existing collections (H. C. Dessauer and M. S. Hafner, eds.). Association of Systematics Collections, Lawrence, 74 pp.

Baker, R. J., and M. Haiduk. 1985. Collections of tissue cultured cell lines suspended by freezing. Acta Zool. Fennica, 170:91-92.

Cato, P. S. 1989. Policies and procedures for the use and management of frozen materials. Texas Cooperative Wildlife Collection, Department of Wildlife & Fisheries Sciences, Texas A&M University, College Station, TX 77843. Unpublished, 114 pp.

———. 1986. Guidelines for managing bird collections. Museology, 7:1-78.

Dessauer, H. C., and M. S. Hafner (eds.). 1984. Collections of frozen tissues: value, management, and directory of existing collections. Association of Systematics Collections, Lawrence, 74 pp.

Dessauer, H. C., and R. A. Menzies. 1984. Stability of macromolecules during long term storage. Pp. 17-20 in Collections of frozen tissues: value management and directory of existing collections (H. C. Dessauer and M. S. Hafner, eds.). Association of Systematics Collections, Lawrence, 74 pp.

Dessauer, H. C., M. S. Hafner, R. M. Zink, and C. J. Cole. 1988. A national program to develop, maintain, and utilize frozen tissue collections for scientific research. ASC Newsletter, 16(2):3-10.

Force, R. W. 1975. Museum collections-access, use, and control. Curator, 18(4):249-255.

Haas, P. 1983. The conservation of photographic collections. Curator, 26(2):89-106.

Hardy, J. W. 1984. Depositing sound specimens. Auk, 101(3):623-624.

Lee, W. L., B. M. Bell, and J. F. Sutton. 1982. Guidelines for acquisition and management of biological specimens. Association of Systematics Collections, Lawrence, 42 pp.

Malaro, M. C. 1979. Collections management policies. Museum News, 58(2):57-61.

————. 1985. A legal primer on managing museum collections. Smithsonian Institution Press, Washington, D.C., 351 pp.

McWilliams, J. 1979. The preservation and restoration of sound recordings. American Association for State and Local History, Nashville, 138 pp.

Porter, D. R. 1986. Developing a collections management manual. Technical Report, American Association for State and Local History, Nashville, 7:1-20.

Pritchard, M. H., and G. O. W. Kruse. 1982. The collection and preservation of animal parasites. Univ. Nebraska Press, Lincoln, 141 pp.

Rempel, S. 1983. Enclosures for housing photographic negatives. Conservation Notes, Materials Conserv. Lab., Texas Memorial Museum, Univ. Texas (Austin),3:1-4.

Ritzenthaler, M. L. 1983. Archives & Manuscripts: Conservation. Society of American Archivists, Chicago, 151 pp.

Ritzenthaler, M. L., G. J. Munoff, and M. S. Long. 1984. Archives & Manuscripts: Administration of photographic collections. Society of American Archivists, Chicago, 173 pp.

Schmidly, D. J., W. R. Barber, P. S. Cato, and M. E. Retzer. 1985. The collection management practices of the Texas Cooperative Wildlife Collection, Texas A&M University. Unpublished, 109 pp. (Available from authors, Department of Wildlife & Fisheries Sciences, Texas A&M University, College Station, TX 77843.)

Simmons, J. E. 1987. Herpetological collecting and collections management. Herpetological Circular, Society for the Study of Amphibians and Reptiles, 16:1-70.

Williams, S. L., R. Laubach, and H. H. Genoways. 1977. A guide to the management of Recent mammal collections. Spec. Pub., Carnegie Museum of Natural History, 4:1-105.

APPENDIX.—*Texas Cooperative Wildlife Collection Policy Concerning the Use and Management of Ancillary Preparations.*

1. STATEMENT OF PURPOSE

The Texas Cooperative Wildlife Collection (TCWC) acts as a repository for materials which document the vertebrate fauna of the world, with primary emphasis on Texas, Mexico, Central and South America. Traditional materials include skins, skulls, skeletons and fluid-preserved materials as well as the field notes and photographs which pertain to these materials. Ancillary materials include phalli, cleared and stained specimens, karyotype preparations, and frozen tissues and cell lines. The TCWC is committed to the proper conservation and management of these materials to maximize their value for scientific research and the advancement of knowledge.

This document serves to establish the TCWC policy concerning ancillary materials for mammal specimens and supplements "The Collection Management Practices of the Texas Cooperative Wildlife Collection, Texas A&M University" (Schmidly, Barber, Cato and Retzer, 1985).

2. ADVISORY COMMITTEE

An Advisory Committee exists to monitor the use and maintenance of ancillary materials in the TCWC. Such a committee was deemed useful to insure equitable access to the materials as well as insure maintenance of professional management standards.

The Advisory Committee for mammalian materials will include the Curator of Mammals, the Assistant Curator of Mammals and Birds, the researcher who deposits ancillary material in the collection, and one faculty member each from the Department of Wildlife & Fisheries Sciences and the Department of Biology.

3. ACQUISITIONS

The TCWC actively acquires ancillary materials to supplement traditional specimen preparations. Ancillary materials are obtained primarily for mammals and reptiles from research in the United States, Mexico, Central and South America. Research is conducted by TCWC staff and researchers in the Department of Wildlife & Fisheries Sciences and Department of Biology through a variety of funding agencies. Materials from other geographic regions may be obtained through exchanges, loans, or gifts. Materials deposited in the TCWC become property of the TCWC.

Acceptance of all ancillary materials for mammalian species is subject to the approval of the Curator of Mammals in consideration of recommendations made by the Advisory Committee. Acceptance of materials will be based on the abundance of the taxon, the geographic source of the material, active research programs at TAMU, requests for materials from other research labs, and the current content of the TCWC ancillary collections. Priority will be given to ancillary material for which traditional voucher specimens are deposited in the TCWC.

Specimens will not be accepted without proof that they were legally obtained and, when applicable, legally exported from the country of origin and imported into the United States. Nor will ancillary material be accepted if restrictions are placed on its use.

4. MANAGEMENT OF ANCILLARY MATERIALS

Record keeping: Ancillary material will be initially numbered and documented by the collectors in the field according to professional collecting standards (Lee et al., 1982; Dessauer & Hafner, 1984; Baker and Haiduk, 1985). Two TAMU laboratories currently maintain records for frozen tissues and karyotype preparations in bound ledgers using separate numbering sequences prefaced by AK or GK. The AK or GK number references the material for future access. Non-frozen ancillary preparations such as phalli are labeled with the preparator's initials and field number.

Upon acceptance of the ancillary materials by the Curator of Mammals, the TCWC staff will be responsible for maintaining all records, for tracking the location of the materials, and for the proper storage and maintenance of the materials.

All ancillary materials will be accessioned by TCWC staff in the Accession File and will receive a TCWC accession number. One TCWC accession number is assigned for a lot of specimens obtained from one source (for example, one field trip or one research project). This step verifies the transfer of materials to the TCWC.

Ancillary materials for mammalian species will be permanently and individually referenced by two methods: 1) TCWC catalog sequence, and/or 2) special numbering sequence (AK, GK, etc.). Ancillary materials which are represented as well by traditionally prepared voucher specimens (skins, skulls, skeletons, or fluid specimens) will receive the TCWC catalog number assigned to the voucher specimen. These materials and their data may be retrieved through either the TCWC catalog number or the AK or GK number. Materials which have an AK or GK number but which are not represented by traditional voucher specimens will not receive additional TCWC catalog numbers; instead, they will continue to be referenced only by the AK or GK number.

This cataloging system was structured to combine the three numbering systems that currently exist (TCWC, AK, and GK) in such a way as to permit easy referencing and retrieval of materials without extensive recataloging. Data concerning all of the materials will be available through the computerized inventory system of the TCWC.

Storage: Ancillary materials will be stored in the TCWC under appropriate storage conditions, The TCWC currently maintains a separate cabinet for storage of phalli, two upright ultracold freezers (-85°C), and one liquid nitrogen unit. Within the storage unit, ancillary materials will be arranged taxonomically to family level, alphabetically for genera and species and numerically within each species.

Use of Specimens: Use of ancillary materials will be monitored by the Advisory Committee to insure equitable access to materials for legitimate research and educational purposes. Specimen use may be categorized into three types, loans, gifts, and exchanges.

Material which is "loaned" will not be consumed by the borrower and will be returned to the TCWC within a specified time period. Examples of this type of material are phalli and karyotype slides. "Gifts" are made of material which will be consumed during analysis, and will not be returned to the TCWC. Most requests for use of frozen tissues will be fulfilled as gifts rather than loans. "Exchanges" of material may be made if the TCWC is designated to receive material comparable in quantity or value to material sent out. Exchanges will be pursued as opportunities to acquire materials for researchers at TAMU and to fill out the scope of taxa or geographic representation in the TCWC.

Restrictions may be placed on the uses of certain ancillary materials upon the recommendation of the Advisory Committee. These restrictions would apply primarily, although not exclusively, to materials taken from rare or endangered taxa and to materials which serve as vouchers for specific research projects. Such restrictions would establish an amount of material which may not be consumed during analysis except under extremely unusual circumstances.

Each request for use of ancillary materials will be considered individually by the five-member Advisory Committee. The recommendation of the Advisory Committee will be forwarded to the Curator of Mammals, who will have final authority on the disposition of material. Each request must include a list of the species requested, the type and quantity of material needed, a short description of the proposed use of the material and the nature of the borrower's research. This information will be reviewed by committee members in view of the rarity of the material requested and the quantity available in the TCWC. The action recommended by the committee will be recorded on a Request for Use form (Fig. 1) which will be sent to the individual requesting to use materials as notification of the action taken.

Provisions for long-term loans will be considered by the Advisory Committee for material collected by an individual which pertains directly to his or her research interests. Once the long-term loan has been implemented by the Curator of Mammals, other uses of this material are subject to the approval of the researcher.

TEXAS COOPERATIVE WILDLIFE COLLECTION
REQUEST FOR USE OF ANCILLARY PREPARATIONS

Request made by: Date:_____
                                             TYPE OF REQUEST:
_____   _____ Gift
_____   _____ Loan
_____   _____ Exchange
_____   _____ Other_____
_____

Material made available by researcher:
        _____ description of material needed
        _____ description of research
        _____ potential use of material
_____
COMMITTEE ACTION -- Request is (accepted, denied, needs further review).

REASONS FOR ACTION:

Request reviewed by:

_____   date_____

_____   date_____

_____   date_____

_____   date_____

_____   date_____
        Curator of Mammals

Fig. 1.—The "Request for Use Form" is to be completed for each request to use ancillary preparations held by the Texas Cooperative Wildlife Collection.

A long-term loan is implemented with the understanding that much of the material will be consumed, but the researcher is requested to follow recommendations by the Advisory Committee concerning limitations on rare taxa and vouchers. Time periods for long-term loans will be negotiated between the researcher and the Advisory Committee.

The TCWC will pay for shipping charges on loans and exchanges, but the recipient will be requested to pay for shipping charges for gifts of materials that must be sent with dry ice. Monies for these charges should be requested by the individual who wishes to use TCWC materials from agencies which are funding his or her research.

Deaccessions: Materials which leave the TCWC as gifts or exchanges will be deaccessioned according to standard TCWC practices. Loan material will not be deaccessioned unless totally consumed during long-term loans.

5. REVISION OF POLICY

This policy is subject to ongoing revisions in order to remain current with professionally accepted standards for the use and maintenance of ancillary materials.

Adopted August 1987

Natural History Museums: Directions for Growth
Paisley S. Cato and Clyde Jones, editors
Texas Tech University Press, Lubbock, 1991, iv+252 pp.

FORWARD INTO THE PAST: A CENTURY OF CHANGE IN VERTEBRATE PALEONTOLOGY COLLECTIONS

Sally Y. Shelton

Abstract.—Progress in the research standards of vertebrate paleontology has not always been matched by progress in collections management standards. Problems that have seriously affected the availability and use of vertebrate fossils range from biological and mineralogical deterioration of individual specimens to outright administrative abandonment of entire collections. The relevance of these collections to evolutionary science has been cast into doubt by some scholars. The need for more and better information from vertebrate fossils has fostered an awareness of the conditions required for their protection. Changing standards and attitudes hold out hope for improved support, management, and use of these collections.

A time of excitement began in 1888 for North American museums. The discoveries of large fossil deposits in the western states spawned an unprecedented collecting frenzy. Expeditions raced to quarry, pack, and ship the largest, newest, or most numerous fossils to museums in the northeastern states. The collections of Othniel Marsh and Edward Drinker Cope, spurred by fierce competition between the two, filled museum exhibits and basements. Vertebrate paleontology advanced rapidly, aided by the volume and diversity of fossils available.

A time of change and confusion exists in 1988. The jumbled booty of the bone wars, some of it yet unpacked and unstudied, remains in museum storage. Vertebrate paleontology is seen by some as one of the most problematic of the systematics disciplines; the relatively small number of specimens (as compared to invertebrate paleontology) and the lack of genetic data (as compared to Recent biota collections) have led some to dismiss the relevance of vertebrate fossils as useful indicators of evolutionary trends and patterns. Collections of vertebrate fossils, ironically, can be studied in more ways and yield more information now than ever before, but some may lose housing and funding before they can be adequately studied.

In 1977, the Advisory Committee for Systematics Resources in Vertebrate Paleontology (an *ad hoc* committee of the Society of Vertebrate Paleontology) reported that ". . . some nationally significant collections are deteriorating as bases for scientific research, and some are actually closing their doors.

These losses are serious, both nationally and internationally, because they damage the very foundation of the science" (Langston *et al.*, 1977). This is still a concern today, and some major collections have been divided or dispersed. Others are in need of secure funding for improved housing and management, hampered by administrative perceptions and priorities.

Vertebrate paleontology is an anomalous branch of systematics: all fossil collections are incomplete in varying degrees, from the loss of soft tissues to the absence of most of the components of an ecosystem. In the words of the 1977 report: "Vertebrate fossil collections comprise one of the smaller factors in the systematics collections universe . . . The science of . . . paleontology is perhaps more dependent upon its systematics collections than is any other branch of the natural sciences. This is because the only way to know extinct [biota] is through their preserved remains; thus the whole body of paleontology rests upon the unique and irreplaceable fossils accumulated in systematics collections since the eighteenth century" (Langston *et al.*, 1977).

The gaps in this body of knowledge are numerous. Bones disarticulate soon after death, attract scavengers, or roll or wash away. Fossilization occurs rarely and unevenly, a chance combination of physical and chemical circumstances that may favor a few groups, or a few individuals, or none at all. Genetic information is lost with the decay of soft tissues, and with it goes the proof of species affiliation required by most systematists working with Recent taxa.

Yet vertebrate paleontology may well be the most public, most highly popularized, and most visible branch of systematics; its immense specimens have immense appeal. It is also one of the most important areas of vertebrate evolution. "The contributions and influence of vertebrate paleontology on the biological sciences is [*sic*] so general and pervasive that specialists in the fields are not always aware of them . . . Paleontology has [,] by defining evolutionary sequences, provided a rational and historical framework for the essentially biological field of systematics" (Langston *et al.*, 1977). Fossil remains are of intrinsic interest to many collectors at an amateur level, but are scientifically meaningful only in the context of time, place, and ecology. This can be established only by the careful acquisition, documentation, and housing of specimens in an organized collection. The management of the specimens and associated data may be the most important and least well understood aspect of paleontological collections.

Vertebrate fossil collections include both large specimens and large numbers of specimens when at all possible. The latter is critically important. Large collections are vital to vertebrate paleontology as the only possible means of gaining insight into the structure and evolution of taxa and ecosystems. Meaningful patterns are drawn only from many specimens.

"Large samples are also essential for all types of growth studies and for other research concerned with ontogenetic development . . . Large paleontology collections . . . provide documentary proof of evolutionary change by sampling many stages in the history of populations of organisms. They also provide standards of the evolutionary development of entire faunas, establishing geological correlation of strata from place to place throughout the world and helping to develop a worldwide timescale for the ordering of events in the history of earth over many millions of years" (Nicholson et al., 1975).

In addition, large fossil collections aid in establishing the range of normal variation within a group. This aids in the refinement and pruning of questionable taxa. Many taxonomic partitions established at the height of the collecting frenzy of the last century have proved to represent differences better attributed to pathology, post-mortem deformation, sex or age polymorphism, or other factors that can be confirmed only by the analysis of many specimens. Evidence from a large series obviates, to some degree, the uncertainty inherent with a lack of genetic material.

The fields that integrally are associated with vertebrate paleontology collections, including biogeography, systematics, stratigraphy, and paleoecology, may not appear to depend on this association. Ideas that begin with meticulous empirical study of individual specimens undergo many levels of synthesis as more evidence from many sources is integrated. The resulting theories may be removed by many levels of abstraction from the specimen-based evidence. An unfortunate consequence of this may be the neglect of collections that are believed to have served their purpose.

Princeton University set a disturbing precedent along these lines when, in 1985, it divested itself of its collections. One reason given was that "the university, if it were to hire another paleontologist, would undoubtedly hire an outstanding theoretician, one who wouldn't need collections . . . someone like George Gaylord Simpson" (Gingerich, 1987). As Gingerich pointed out, this would have been news to Simpson, who acknowledged the importance of his work in collections, and who began his long career "collecting in the basement" of the Peabody Museum (Simpson, 1978).

By many standards, vertebrate paleontology is not an experimental science. Conclusions drawn from one set of specimens, or a single specimen or collection, may not be replicable in other collections if truly comparable material does not exist or is not available. Each specimen is the result of unique taphonomic and geochemical events. These factors may affect adjacent specimens or even adjacent elements of a single specimen, quite differently. Many specimens are also unique taxonomically, existing in the profoundest possible isolation until such a time as others may be found. Specimens are constantly being reexamined in the light of new evidence and new ideas, and often yield unsuspected data when analyzed in new ways. The uniqueness of the specimens and their unknown information mandates their care and militates against the idea that they are less important than the current ideas drawn from them.

Stephen Jay Gould wrote: "Contrary to the romantic image of science and exploration, many important discoveries are made in museum drawers, not under adverse conditions in the parched Gobi or the freezing Antarctic. And so it must be, for the nineteenth century was the great age of collecting—and leading practitioners shoveled up material by the ton, dumped it in museum drawers, and never looked at it again" (Gould, 1985).There is still, however, a popular tendency to think of paleontologists as stamp collectors, and to assume that the great wave of collecting has yielded all the answers that can be drawn from specimens. One standard museology text dismisses the significance of collections as sources of future information. "In the latter half of the nineteenth century, collecting by [natural history museums] and the study of these collections gave great impetus to the advance of scientific knowledge . . . Museum research has declined in importance because it must deal chiefly with the static description and classification of things. *That work is now largely done* [emphasis added] except for dotting 'i's' and crossing 't's'" (Burcaw, 1975).

How can two such disparate views of the continuing worth and relevance of collections be reconciled? In a time of straitened budgets and space, administrators may not be willing to gamble on the future uses of inactive collections.

Conservation and management of vertebrate fossil collections is important if they are to be of future use. Too often, fossils are assumed to be inherently stable; specimens that have lasted thousands or millions of years *in situ,* it is reasoned, should last indefinitely in any kind of museum storage or exhibit with no further care. In truth, uncontrolled microenvironmental conditions often combine with the effects of unprecedented exposure to break down both the organic and inorganic constituents of fossils relatively quickly. From the familiar concentric cracking of ivory to the ravages of pyrite disease, fossil deterioration has been noted as a subject of serious concern for many years. The research on the causes of, and treatment for, these problems is emerging only now as a field in its own right, as scientists begin to analyze the biological and geological nature of fossilized material and to view specimens as being in constant chemical interaction with their environments.

Housing and management of the collections requires adequate support, staffing, and training. To survive in a time of limited resources and to secure these requirements, vertebrate paleontology collections must be demonstrably relevant to the interests and goals of their institutions. It is important that the usefulness of these collections be known outside the paleontological community. Many collections are held in public trust; their importance, through education at all levels, must be communicated to the public. Vertebrate paleontology generates strong interest by its very nature and presents an opportunity for effective education on the value of natural resources and the need for their protection.

In New Mexico, a group of "paleontologists, state legislators and laymen" collaborated in a remarkably successful effort to secure legislative protection and financial support for the study and collection of New Mexico's valuable paleontology resources (Rowe *et al.,* 1982). A survey of these resources generated public concern, which in turn engendered protective legislation. In 1980, a bill established the New Mexico Museum of Natural History, with a dedicated program in vertebrate paleontology.

For scholars, the relevance of vertebrate paleontology collections to the study of evolution has been cast into doubt by some recent studies in cladism. Gauthier *et al.* (1988) reaffirmed the importance of these collections by demonstrating that it is not possible to derive the same relationships between extant vertebrate groups with and without fossil evidence. In particular, the position of mammals in amniote phylogeny varies markedly.

"Some biologists have belittled the role of fossils in phylogenetic inference, because they are only parts of organisms. Missing data can be a problem, but it is not one to which extant taxa are immune. Clearly, extant taxa can offer a greater number and variety of characters than can extinct taxa. But this does not justify equating the mere amount of information with its relevance. A character can be germane to phylogenetic questions at one level of analysis and not so at another . . . We contend that in a particular case, the phylogenetic position of mammals among amniotes, evidence from fossils is of greater relevance than the phenotypic diversity available only from Recent taxa. Even without such an example, we would advise systematists to avoid prejudging the importance of one kind of organism over another, in effect thinking phenotypically, and consider all of the evidence regardless of its nature" (Gauthier *et al.*, 1988).

The relevance of vertebrate paleontology collections to public and to research interests is clear; the management of these collections must be ensured. The collections manager occupies a relatively new niche in collections hierarchies, taking on the concerns of collections care and use as curators focus increasingly on other lines of research. Collections improvement grants, such as those administered by the National Science Foundation, attempt to provide some collections needs. New avenues of data management and nondestructive testing are being developed and refined. Some collections are still dealing with problems caused by neglect or misuse, deterioration of chemicals used as consolidants or adhesives, and information loss. The lack of adequate space is a universal concern, as is the lack of adequate staff with appropriate and ongoing education in collections science.

If vertebrate paleontology collections are lost to neglect, misunderstanding, lack of support, or preventable deterioration, no replacement can ever be adequately made. These holdings are of critical importance, in ways that the bone warriors knew and in ways that they could not have foreseen. Vertebrate paleontology cannot move forward into the past by leaving its collections behind.

ACKNOWLEDGMENTS

Special thanks are due to Timothy Rowe, Wann Langston, Jr., Ernest Lundelius, Jr., and John A. Wilson, all of the Vertebrate Paleontology Laboratory (VPL), Texas Memorial Museum, for their helpful critiques and suggestions. Thanks are also due to Rickard Toomey and John Buckley of the VPL. I also wish to thank the Collections Care Pilot Training Project, its sponsor, the Bay Foundation, and the coordinators of and participants in the CCPTP, Natural History Museum of Los Angeles County, for their support and vision. This is publication number NS-11 of the Texas Memorial Museum.

LITERATURE CITED

Burcaw, G. E. 1975. Introduction to museum work. The American Association for State and Local History, Nashville, 192 pp.

Gauthier, J., A. G. Kluge, and T. Rowe. 1988. Amniote phylogeny and the importance of fossils. Cladistics 4:105-209.

Gingerich, P. D. 1987. Simpson as a model. Palaios 2:2.

Gould, S. J. 1985. The flamingo's smile: reflections in natural history. W. W. Norton & Company, New York, 476 pp.

Langston, W., et al. 1977. Fossil vertebrates in the United States: the next ten years. Report of the Society of Vertebrate Paleontology Advisory Committee for Systematics Resources in Vertebrate Paleontology, Austin, 56 pp.

Nicholson, T. D., B. Schaeffer, T. Galusha, M. C. McKenna, M. F. Skinner, B. E. Taylor, and R. H. Tedford. 1975. The fossil mammal collections of the American Museum of Natural History. Curator 18: 16-38.

Rowe, T., R. Cifelli, and B. Kues. 1982. The instrumental role of paleontology in the funding and development of a major new museum. J. Paleontology 56: 839-841.

Simpson, G. G. 1978. Concession to the improbable: an unconventional autobiography. Yale University Press, Yale University, 280 pp.

Natural History Museums: Directions for Growth
Paisley S. Cato and Clyde Jones, editors
Texas Tech University Press, Lubbock, 1991, iv+252 pp.

DESTRUCTIVE ANALYSIS OF ARCHAEOLOGICAL COLLECTIONS: BETWEEN SCYLLA AND CHRYBDIS

Allen S. Bohnert and Margo Surovik-Bohnert

Abstract.—Anthropology collections, including those obtained through archaeological methods, historically have been considered a componet of natural history museums. We will examine what has been described as a "crisis in archaeology", and relate it to museum collection management. Specifically, the need for guidelines governing destructive analysis of archaeological collections will be discussed and recommendations for such a policy will be presented.

During the past two decades there has been an increased awareness on the part of the American public, lawmakers, archaeologists, and others that our nonrenewable archaeological resources are being lost at an alarming rate. Our archaeological heritage is finite for any given time period and we must "conserve and manage this finite set of resources to ensure its best use over a maximum length of time" (McGimsey and Davis, 1977).

In the 1970s, recognition of the accelerating rate of attrition of the archeological resource base prompted the publication of reports describing a crisis in American archaeology (Davis, 1972; McGimsey and Davis, 1977). During this period, Lipe (1974*b*) called for the archeological community to shift from a salvage mode to a "conservation model." Conservation archeology is based, in part, on an ethic that calls for the preservation of the resource base as a whole and for treating archaeological "salvage as the last resort—to be undertaken only after other avenues of protecting the resource have failed" (Lipe, 1974*b*). Much of the concern and the subsequent efforts related to resolving the crisis in archeology focused on the preservation of archeological sites, as the primary resource.

However, the archaeological resource base also includes all artifacts, specimens, and documentation resulting from an archaeological investigation. It became apparent that the crisis in archaeology included a crisis in the curation of archaeological collections (Ford, 1977; Heritage Conservation and Recreation Service, 1980). "As early as 1975 a consensus had emerged that there was a problem, [and] that it would get worse" (Marquardt *et al.*, 1982). Discussions related to the crisis in archaeological curation commonly have revolved around a lack of accountability, access, and long-term preservation of archaeological collections. The conservation of archaeological collections has also become an important issue (Bourque et al., 1980; Morris, 1981; Raphael et al., 1981). Conservation has begun to play an integral role in all phases of archaeological research, particularly in the areas related to collection use.

When one examines the concept of wise use of a finite resource over a maximum length of time, concerns about long-term preservation of archaeological collections arise. This is especially pertinent when considering the analytical techniques used in studying archaeological collections. A wide variety of analytical techniques may be employed, ranging from the "simple" cleaning ceramic studies to more specialized techniques involving trace element analysis of certain materials. Many techniques cause the destruction of portions of objects and, perhaps more important, of data that may be useful to future researchers when analyzing archaeological collections.

The issues concerning the crisis in archaeology and archaeological curation have a direct relationship to policies governing collection research. It is helpful to illustrate some of the idiosyncracies of archaeological collections, along with recent developments in the field, to better understand the relationships among long-term preservation, wise use of collections, and research policies.

The size and the scope of archaeological collections has increased dramatically in the past two decades due to several factors. The passage of certain federal legislation and other initiatives such as the National Historic Preservation Act of 1966 as ammended in 1980 (U.S. Government, 1966, 1980), the National Environmental Policy Act of 1969 (U.S. Government, 1969), the Archeological and Historic Preservation Act of 1969 as ammended (U.S. Government, 1980), and the Archeological and Historic Preservation Act of 1974 (U.S. Government, 1974) resulted in archaeological collection growth.

Archeological method and theory also have changed, with a "greater emphasis being placed upon processual interpretation, quantitative analysis and interdisciplinary efforts" (Bell, 1985). Thus nearly all cultural material and a significant sample of noncultural matrix recovered from an archaeological investigation is retained. The present rationale for keeping all relevant or analyzed materials, rather than following the previous practice of discarding all but a sample, is based on several premises. Two of these are verification of research results and availability of the material for future research. The scientific method, to which archaeological method and theory ascribe, requires replicateable tests and verifiable conclusions. Thus the objects, material samples, and associated documentation related to any analysis must be properly curated.

Until relatively recently, the documentation associated with an archaeological investigation ". . . has been viewed as *private property* to be kept by the archaeologist who excavated a site or surveyed an area. As long as the notes are in the archaeologist's possession, it signifies an intention to write up the site or survey, even if a score of years has elapsed" (Ford, 1984). Frequently the documentation never was united with the collection creating "one of the major problems with using previously excavated collections in many museums or repositories" (Towle and MacMahon, 1987). Today the Secretary of the Interior's Standards for Archeological Documentation for federal collections (U.S. Government, 1983a, 1983b) and other federal regulations (U.S. Government, 1987a, 1987b) require the preservation and availability of not only the materials collected but any associated documentation for all federal collections. The Society of Professional Archaeologists, stresses similar requirements (McGimsey and Davis, 1977).

Additionally, there has been an increased interest in historic archaeology, and as Salwin (1981) stated, "The volume of curateable material recovered from even a modestly sized late 19th or 20th century excavation is truly wondrous to behold." As recent archaeological collections grew in size and scope, so grew the magnitude of the curatorial workload. Generally speaking, there has not been a similar increase in the support for curatorial functions, placing the collections at risk and limiting access to them. There is, however, an increasing awareness of the need to save our archaeological resource base for future investigations (Green, 1984; U.S. Congress, Office of Technology Assessment, 1986).

A common theme in nearly all discussions concerning curation of archaeological collections is accessibility for future research. It has been proposed that the "use of existing archeological collections in repositories is recommended in preference to the field recovery of new collections, for research, exhibition, and educational purposes" (Heritage Conservation and Recreation Service, 1980). As yet archaeologists are not fully exploring the research potential of previously recovered collections (Plog, 1980). One can envision a time in "the not-too-distant future [when] most of America's archaeological resources will be in museums, not in the ground. With the sites gone, it is imperative that the collections survive" (Ford, 1984). Consequently, "more and more archaeological research will have to take place in museums and other facilities where collections are curated" (Christensen, 1979).

Preparation for such research, including provisions governing destructive analysis, is essential if museums are to accommodate researchers. This preparation necessarily involves establishing policies and procedures governing collection research. It is imperative that museums, universities, and other repositories holding archaeological collections develop policies to guide decisions concerning approval for destructive analysis. The establishment of such a policy is critical to the management of any type of museum collection. Management includes balancing diverse, sometimes conflicting priorities, and "good management could include the destruction of a resource" (Roberts, 1983) under appropriate circumstances.

In concurrence with the principles of conservation archaeology, current cultural resource management and traditional museum collection management concerns, the long-term preservation of archaeological collections must be considered prior to authorizing any use. The collections are limited and destructive analysis irrevocably erodes the resource. Conversely, a number of research questions cannot be pursued unless some consumptive techniques are permitted. Also, certain archaeological material samples are recovered solely for the purpose of destructive analysis. It can be argued that a "research collection gains more importance as it is studied because more of its potential is realized for future work" (Ford, 1977).

The goal of any policy guiding destructive analysis should be to reduce the negative impacts of destructive analysis, while maximizing the benefits of such analysis on any nonrenewable collection. Achieving a balance between excessive control and flagrant resource exploitation is difficult. A carefully devised policy governing research, especially that including destructive analysis, is essential to finding a resolution to this dilema.

It is recommended that institutions housing research collections develop and implement policies and procedures to evaluate, guide, limit, and enable destructive research. A policy statement covering destructive analysis should be provided to each researcher. Certain institutions, including the Museum of Indian Arts and Culture/Laboratory of Anthropology (1988), National Museum of Natural History (1988), Peabody Museum of Archaeology and Ethnology (1988) and The University Museum (1988), have developed such policies. A review of these policies along with our collection management experience suggests that the policy should address and clarify the following.

1. What are the scheduling requirements for collection access?
2. Who will review submitted proposals (perferably a collections committee, a conservator and, if circumstances warrant, appropriate scientists from other institutions)? An important point to make here is that regulations or ethical considerations may mandate consultation with appropriate ethnic or other concerned groups when certain types of materials are being studied.
3. The importance of maintaining and respecting the integrity of each object.
4. Who will ultimately determine where and how samples may be removed from an object (usually the curator or collections manager)? The curator or conservator will usually oversee, and sometimes actually perform, the sampling.
5. Who will perform restoration treatment on an object (usually only the institution staff) and who will pay for it (the researcher requesting the test)?
6. When is retesting of an object permitted?
7. What are the reporting requirements, including object documentation and research data?
8. What happens to samples or portions of samples remaining after the testing (these should be labeled and returned to the institution with a report of findings whether or not formal publication results)?

9. Who will be held financially responsible for any damage to the object beyond the approved test area (the researcher requesting the testing)?

10. The institution may require that the testing be done on the premises. If this is not feasible, then, if warranted, the object should be accompanied by a staff member who would be present during the testing.

11. What are the recommended standard handling and conservation procedures?

12. If objects leave the premises, what are the prescribed loan procedures that must be followed?

13. If an object in the custody of the institution is the property of another party (for example, federal collections at a state repository), permission from the legal owner to conduct the test may also be required.

Institutions should develop and utilize a Request-to-do-Destructive-Research form or a similar document that researchers must complete. The form should include, at a minimum, requests for the following data: the researcher's qualifications; a description of the research or testing methods; where and when the testing will be done; who will do the testing and who will do the sampling; the size (measurement) of the sample needed for the testing; the proposed location on the object of the sample removal site; number of samples needed per object; number of objects to test; list of actual objects to test, if known; purpose and importance of the tests (expected results of the tests); finally, whether or not the testing could be done using nondestructive methods. The form should also request personal references, curriculum vitae, pertinent bibliography, and a copy of the official research proposal or research design. The research design for a given project may already include much of the information requested.

Research proposals should be carefully evaluated and the following criteria are recommended for consideration: (1) the proposal has merit and is reasonable; (2) the researcher has a sound "track record" and publishes results from research projects; (3) the likelihood the proposed analytical methods will yield the intended results; (4) the researcher or analyst is qualified to conduct the study; (5) the amount and the size of the samples and the number of objects to be tested is actually necessary to obtain the results; (6) each object can be tested without incurring significant damage or forfeiture of future use; and (7) the use of the collections is justified.

The institution's research proposal review committee should be thoroughly comfortable with a proposal, and not hesitate to obtain additional information on a project prior to approving destructive testing. The committee could also suggest alternative techniques or a modification of the proposed methods, if deemed necessary. It is highly recommended that in-house staff members also submit requests to perform testing on objects and these requests should be evaluated by the review committee. Conservation treatments and routine use also should be examined to determine the long-term effects on future research.

Institutions must ensure that they receive at least one copy of all research results and publications within a reasonable time frame upon the completion of the research. The institution should be given a credit line in all printed materials related to that institutions' objects. Researchers should sign a statement that will bind them to the above mentioned commitments and to the conditions noted in the policy statement. Additional testing should not be allowed unless the researcher has complied with the research stipulations.

Some thought needs to be given to the scope of material available for destructive analysis. Complete or whole objects should be avoided. Type specimens should rarely, if ever, be sampled. Destructive analysis could be focused on those specimens that were collected for that purpose or are of little value for other types of research, redundant, unprovienced, intended for exchange, and fragmentary. The value of all proposed research must outweigh the damage to the object.

The data collected through research and testing are important. Adequate samples are needed for later analysis and a data base developed that will hold answers to questions that are as yet unformulated. Lipe (1974b) stated over a decade and a half ago: "If our field is worth a damn, what we think is significant today will not be significant twenty or fifty years from now because the field will have changed." We need to obtain the maximum benefit from samples taken; collect information that might not be specifically pertinent to the research question, but would be useful to other researchers in the future.

We can never totally resolve the preservation-use dilemma. We can, however, develop effective collection management processes that enhance the positive, while lessening the negative aspects of various destructive analysis techniques. Museums are beginning to do this with the emergence of formal collection use and testing procedures, but more must be done. We must invest in and lend support to studies that evaluate scientific testing and data recovery methods that endeavor to create new, less destructive, means of accomplishing certain research. More thought must be given to obtaining data recovery through nonconsumptive methods and to modifying techniques to minimize the damage to museum collections. "In the future we must pursue increased sophistication in scientific learning strategies" (Binford, 1986) if archaeology is to continue to progress and if we are to continue to receive the support of the American public.

The consequences of 'business as usual' will be continued resource attrition and an inability to conduct future research.

ACKNOWLEDGMENTS

We would like to thank Steven Chomko and Ann Johnson for their constructive review of the paper. All errors of omission and comission are, however, our own.

LITERATURE CITED

Bell, J. 1985. Integrating archaeological collections into museum collections. Paper presented at the 18th Annual Meeting of the Society for Historical Archaeology, Boston.

Bourque, B. J., S. W. Brooke, R. Kley, and K. Morris. 1980. Conservation in archaeology: moving toward closer cooperation. American Antiquity, 45:794-799.

Binford, L. R. 1986. In pursuit of the future. Pp. 459-479, in American archaeology past and future (D. D. Meltzer, D. D. Fowler and J. A. Sabloff eds.). Smithsonian Inst. Press, Washington, D.C. and London, 479 pp.

Christenson, A. L. 1979. The role of museums in cultural resource management. American Antiquity, 44:161-163.

Davis, H. 1972. The crisis in American archeology. Science, 176:267-272.

Ford, R. I. 1977. Systematic research collections in anthropology: an irreplacable national resource. Harvard University, Peabody Museum of American Archeology and Ethnology, Cambridge, Mass, 81 pp.

———. 1984. Ethics and the museum archaeologist. Pp 133-142, in Ethics and values in archaeology (E. L. Green, ed.). The Free Press, New York, 301 pp.

Green, E. L., ed. 1984. Ethics and values in archaeology. The Free Press, New York, 301 pp.

Heritage Conservation and Recreation Service. 1980. The curation and management of archeological collections: a pilot study. Heritage Conservation and Recreation Service, Washington, D.C., 140 pp.

Lindsay, A. J., Jr. 1980. Artifacts, documents, and data: a new frontier for American archaeology. Curator, 23:19-29.

Lipe, W. D. 1974*a*. A conservation model for American archaeology. The Kiva, 39:213-245.

———. 1974*b*. General discussion of McGimsey, the restructuring of a profession. Pp 189-190, *in* Proceedings of the 1974 Cultural Resource Management Conference, Federal Center, Denver, Colorado (W. D. Lipe and A. J. Lindsay Jr., eds.). Museum of North Arizona Technical Series, No. 14, 205 pp.

Marquardt, W. H., A. Montet-White, and S. C. Scholtz. 1982. Resolving the crisis in archaeological curation. American Antiquity, 47:409-418.

McGimsey, C. R., IV. and H. A. Davis, eds. 1977. The management of archeological resources, The Airlie House Report. Society for American Archaeology Special Publication, Washington, D. C., 124 pp.

Morris, K. 1981. Conservation of archaeological collections. North American Archaeologist, 2: 131-138.

Museum of Indian Arts and Culture/Laboratory of Anthropology. 1988. Request for permission to do scientific testing. Unpublished document. Museum of New Mexico, Santa Fe, NM.

National Museum of Natural History. 1988. Anthropology sampling committee policies, procedures, and guidelines (DRAFT). Unpublished document. Smithsonian Institution, Washington, D. C.

Peabody Museum of Archaeology and Ethnology. 1988. Permission to conduct destructive analysis and general information for outgoing loans. Unpublished document. Harvard University, Cambridge, MA.

Plog, F. 1980. The ethics of archeology and the ethics of contracting. Contract Abstracts and CRM Archeology, 1:10-12.

Raphael, B., K. Laitner, M. Surovik-Bohnert, and D. Scott. 1981. Field conservation manual. Appendix B, *in* Archeological Resources in Southwestern Colorado, Cultural Resources Series No. 1. Bureau of Land Management.

Roberts, M. 1983. Discussant remarks, management and preservation. American Archeology, 3:228-229.

Salwen, B. 1981. Collecting now for future research. Pp. 567-573, *in* The research potential of anthropological museum collections (Anne-Marie Cantwell, James B. Griffin, and Nan R. Rothschild, eds.). Annals of the New York Academy of Sciences, Vol. 376. The New York Academy of Sciences, New York.

Towle, L. A., and D. A. MacMahon, eds. 1987. Archeological collections management at Minute Man National Historical Park, Vol. 1. National Park Service, Boston, MA, 344 pp.

United States Congress, Office of Technology Assessment. 1986. Technologies for prehistoric and historic preservation, OTA-E-319. U.S. Government Printing Office, Washington, D. C., 198 pp.

United States Government. 1966. An Act to Establish a Program for the Preservation of Additional Historic Properties Throughout the Nation and for Other Purposes. P. L. 89-665; Stat. 915; 16 U. S. Code 470.

———. 1969. National Environmental Policy Act. P. L. 90-190; 91; 16 U. S. Code 470.

———. 1974. An Act to Ammend the Act of June 27, 1960, Relating to the Preservation of Historical and Archaeological Data. P. L. 93-291; 88 Stat. 174; 16 U. S. Code 469.

———. 1980. National Historic Preservation Act Ammendments. P. L. 96-515; 94 Stat. 2987; 16 U. S. Code 470.

———. 1983a. Archeology and historic preservation; Secretary of the Interior's standards and guidelines. Federal Register, 48(190):44716-44742.

———. 1983b. Code of Federal Regulations 36 CFR 66, Recovery of scientific, prehistoric, historic, and archeological data: methods, standards, and reporting requirements. Federal Register, 48(190):44716-44742.

———. 1987a. Code of Federal Regulations 36 CFR 79, Curation of federally-owned and administered archeological collections; proposed rule. Federal Register, 52(167):32640.

———. 1987b. United States Government, Code of Federal Regulations, National Park Service, Native American Relationships Management Policy. Federal Register, 52(183):35874-35878.

The University Museum. 1987. Destructive testing policy and procedures. Unpublished document. University of Pennsylvania, Philadelphia, PA.

Exhibits and Education

Natural History Museums: Directions for Growth
Paisley S. Cato and Clyde Jones, editors
Texas Tech University Press, Lubbock, 1991, iv+252 pp.

THE EVOLUTION OF EXHIBITIONS
IN A NATURAL HISTORY MUSEUM

Louise Lauretano DeMars

Abstract.—The foundation of natural history museums is the collections, the real objects upon which the museum's research, education, and exhibitions ultimately are based. The exhibitions are the magnet that has kept the public coming back to our institutions generation after generation in increasing numbers. For the general public, principal access to the collections is through exhibits, which are an amalgam of science, educational planning, and design. Exhibitions are the vehicle through which the museum places the observer in contact with carefully selected objects by creating a setting conducive to a personal response, either intellectual or emotional, on the part of the observer. The interdisciplinary nature of our material and the growing sophistication of our audience has put new demands on natural history museum professionals to work as a team, by formalizing the exhibit development process, to create exhibitions that are scientifically sound, educationally stimulating, architecturally engaging, and entertaining. In the natural history museum setting, refinement of the exhibition development and design process, new technology in production methods, a growing interest in learning theory, concern about conservation, and the evaluation of audience responses have contributed to the professionalization of exhibition people and have heightened their determination to achieve excellence in the production of exhibitions.

The foundation of natural history museums is the collections, the real objects upon which the museum's research, education, and exhibitions ultimately are based. The exhibitions are the magnet that has kept the public coming back to our institutions generation after generation in increasing numbers.

The 28 March 1988 issue of *Newsweek* magazine published a Louis Harris and Associates poll on how Americans use their leisure time. It stated that museum attendance was up by 24 percent since 1984, an increase second only to VCR sales.

For the general public, principal access to a museum's collections is through exhibits, which are an amalgam of science, education concepts, and design. Exhibitions are the vehicle through which the museum places the observer in contact with carefully selected objects by creating a setting conducive to a personal response, either intellectual or emotional, on the part of the observer. Reverberations of a successfully completed exhibit affect the entire museum family both in terms of morale and finances. From the shipping and receiving clerk to the guards in the museum proper, everyone is touched by an exhibition production in some way.   An exhibition should be viewed as a museum project, not as an isolated exhibit project.

When the present Peabody Museum building opened in 1925, there was no such thing as an exhibit designer. The curators, assisted by technicians and preparators, installed the specimen material and label text, and created the first exhibits for the museum. Thirty-five years later, in 1960, in the evolution of the exhibit process, a formal exhibit department was established and for the first time design became a factor in the process.

I have worked in the exhibition field at the Yale Peabody Museum of Natural History for 25 years. Over those years, through seven administrations, the process by which we have developed exhibitions has changed dramatically depending on the level of public-side interest on the part of each director. During my early years, the curator and designer worked out the exhibit concepts, took the project, as summarized on paper, through committees for approval, and then, when production was ready to begin, other staff members were called in to do their jobs. Many exhibits were developed primarily to appeal to the few Yale students who may or may not have bothered to view them. The level of scrutiny by other individuals during the development of an exhibition was varied and mostly limited to scientists. The focus was on the specimen, conservation, how much text we could get into any given area and, of course, meeting the deadline for opening night.

In 1978 during one of our more progressive, public-oriented administrations, while we were immersed in a particularly hectic temporary exhibition schedule, I kept thinking, "There has got to be a better way." Ideas began to take form about a different method of developing and producing exhibitions. Our lack of consideration for the public—our true audience—the absence of goals for each exhibit, false starts, changes in midstream, all contributed to the hatching of our new approach to exhibition development, production, and evaluation. The process by which exhibitions are produced at the Peabody Museum of Natural History has evolved beyond that of the curator and designer creating displays in a vacuum with no input from other areas of expertise.

There are three areas that should be explored further and refined when considering future successful natural history museum exhibitions: the first area is the team approach to exhibition development, production, and evaluation (and for the purpose of this paper, I shall limit my remarks primarily to development); the second area is to find new ways to create exhibitions with multiple learning levels; the third is our concern for professional standards within the exhibit profession.

TEAM APPROACH TO EXHIBITION DEVELOPMENT, PRODUCTION, AND EVALUATION

Several factors have put new demands on natural history museum professionals to work as a team. These elements are the interdisciplinary nature of our natural history material, the growing media sophistication of our audience, and our recognition that many areas of expertise contribute to successful exhibitions. To create exhibitions that are scientifically sound, educationally stimulating, architecturally engaging, and entertaining, we must work as a team, by sharing, communicating, and respecting each other's professionalism.

The team procedure we have developed for the Peabody Museum has undergone constant change and may not suit the needs of other institutions. The unique qualities and conditions inherent in your institution require that you embrace the team-approach philosophy and develop a procedure to suit your institution. "Unique qualities and conditions inherent in each institution" simply means anything from mere eccentricity to outright weirdness. I have found four ingredients in a successful exhibition production, they are: philosophy, people, procedure, and psychology. The team approach to exhibition development, production, and evaluation is made up of all four.

Philosophy

The team approach formalizes interactions among the team members and builds a positive approach through the sharing of knowledge. Each team member is asked not just to do a specific job, but to buy into the project, bringing in his or her professional expertise. Members are asked to contribute their expertise on an equal footing with other team members in the early phases of project planning and exhibition development. Through this early sharing of knowledge and information, the content of the exhibition and realistic goals and objectives are established. Throughout the course of the development phase, different core team members take lead roles at different stages of development, but always in consultation with other team members. Team development creates a situation of shared responsibility among the team members and does not allow the project to become totally dependent on any one individual.

People

The people fall into two, but not separate, categories, the exhibition development team members and the audience. The first category may provide the most diversity from institution to institution, with staff size and scope offering unique constraints and opportunities. The core team consists of one or more curators, a coordinator, a designer, and a public educator.

The curator is responsible for the scientific integrity of the exhibition. He or she determines and provides the appropriate collection material and writes the scientific content appropriate to each stage of development, in consultation with the other core team members. The coordinator acts as the facilitator and is responsible for overseeing the organization and management of the project, integrating all aspects of the project into a working schedule and monitoring all activities of the entire exhibit team. The designer, in consultation with the other team members, renders the theme of the exhibition into an architectural reality. The public educator serves as an advisor by knowing the museum's audience. Through the formative stages of the exhibit, the public educator continually evaluates the project to see that it develops in terms the public can appreciate. The public educator also oversees lay-reader input into the development process by finding individuals who have little knowledge of the subject matter, are willing to review scripts and designs, and are able to comment on them as the project progresses. Including the public educator in the development process was one of the innovative aspects of our team procedure.

Before the design phase begins, the team expands to include the construction supervisor, the conservator, the registrar, the preparators, and other personnel, depending upon what staff you have to draw on and the demands of the exhibition. This expansion takes place in order for these people to have the opportunity to review the project, to become aware of what their role will be in the project, and to determine the needs of their special contribution. In this way, both the management of the project and the design and content of the exhibition reflect the early input of all these specialists. By drawing upon the expertise of such groups or individuals early in the project, you eliminate later surprises and glitches.

The audience participates in this development process by providing the team with information during predevelopment audience testing, through development of lay-reader input and through a postexhibition opening evaluation process.

Procedure

The procedure for development is a series of six steps that brings the project from original proposal to a project ready for production with built-in checkpoints, reviews, evaluations, and approvals. Who checks, reviews, evaluates, and approves may vary depending on the makeup of your organization. It is important, if you decide to develop a team approach to exhibition development, that you remain flexible enough to review and evaluate both the management and production of each project and the completed exhibitions, and make adjustments to your procedure as needed.

Step 1. The proposal.—This is a form containing a brief statement of the objectives, the exhibit concepts, and a brief list of potential collection material. In our institution the idea for an exhibit proposal may be submitted by anyone but must be signed by a curator, which is his or her commitment to provide the scientific input for the project. One of our "weirdnesses" is that the individuals making up the team are all full-time, paid museum staff, except for the lay-readers and the curators. The curators are Yale faculty members who, because of their special knowledge relative to the collections, have been appointed as curators of our institution. However, they do not get paid for being a curator.

Step 2. The theme statement.—This is an expanded statement on the exhibit concepts and objectives and also indicates the intended audience. It is produced by the curator in consultation with the other core team members. The theme statement includes the proposed title of the exhibit, a listing and brief description of the available specimen material, and information about loan material, including conservation needs. The theme statement also includes a plan-view of the proposed exhibit location, and an exhibit strategy showing the major topics to be covered and in what order—sequential or random, point of emphasis, story line, main units, and the breadth of coverage. The purpose of the theme statement is to provide the core team with a document that will allow them to begin to form concepts of how the new exhibit will be developed. It is also the stabilizing document that keeps subsequent planning in line with the original idea.

Step 3. The initial phase-first design phase.—The purpose of this step is to permit an initial evaluation of the feasibility of the design concept in relationship to the first draft script. The core team works together, each member taking the specific role needed to bring the project to this phase. The design is created in scale conceptual model form accompanied by two-dimensional scale elevations of each exhibit section that are keyed to a corresponding rough draft script and specimen list. Through the public educator, the lay-readers' comments are presented at this time.

Step 4. The initial time and cost estimate.—This is the stage that will grab the attention of your business manager. Input from all team members is necessary to develop this document, because each determines a time-and-cost analysis for his or her part of the project. This information is provided to the coordinator for preparation of a preliminary project time-and-cost estimate.

Step 5. The second design stage-second draft script.—This design phase consists of scale elevations and floor plans and is accompanied by a final scale model and draft script. The purpose of this step is to permit a final evaluation of the theme, concepts, script, and design—a process that allows the team to fine-tune the project before major museum resources are committed to the production phase.

Step 6. The final specifications stage.—The final script, together with design and construction specifications, is now the focus of attention. The final cost estimate, time analysis, and production schedules are determined at this time.

Psychology

Interesting things happen when you try to effect change. When I first proposed the idea for a team approach in 1978, I was told "not to make waves." I had plenty of time to think about it, because in reality I was unable to pursue this concept until a change of administrations in 1982. By that time, I had accumulated enough information from my Peabody colleagues to have a solid idea of how we should approach developing our procedure—by a team approach, of course.

The majority of staff involved in the exhibition process was eager to help develop this new procedure, but not all were convinced that we should try something new. Feelings ran high and some people felt threatened by a procedure that would allow all participating team members to scrutinize each individual's area of expertise. Some, and I stress some, of the curators to this day feel that the team procedure takes the power and control out of the hands of the curator and one curator is unwilling to participate in a team effort, insisting that only the curator and designer work out the details of the exhibit. It became apparent that the participating professional staff of the museum did not enjoy the same respect from the academic staff as they had given to the latter group. The same curators do not acknowledge the diversity of our audience, rather viewing us as a traditional university natural history museum. But in spite of that, when the team approach is used, a wonderful sharing, camaraderie, and communication has come from this experience. When all systems are go, the end product reflects that effort and our audience responds.

So, in the evolution of the exhibition process, first we had the real object, which has always been the focus of the natural history museum exhibit. We then added conservation of the object as a major concern. More recently we have pursued team development and have recognized the importance of evaluation. And, for the first time, added to the importance and concern for the real object came the awareness of the need of our true audience—the public. We became "people-conscious." What we exhibit did not change, but how we exhibit it did.

MULTIPLE LEARNING LEVELS

Just when we thought we knew what we were doing, in June 1988, Professor Howard Gardner of the Harvard Graduate School of Education told us during his American Association of Museums presentation, entitled "Designing for Multiple Frames of Mind," that seven ways in which people learn have been identified. He told us that learning theory is critical when determining the content of exhibitions—and who said that this job was going to be easy? The evolution continues; this gives museum exhibit professionals a new avenue to explore in an attempt to produce exhibitions that have more appeal and reach a broader audience.

PROVIDING INFORMATION TO OUR AUDIENCE

In the past, the environments created by exhibits in natural history museums were a little like church and a lot like libraries, both in the hushed atmosphere and, at times, in the attitude toward the content. One experienced a quiet reverence or a communion with the real objects. Exhibits today are developed and designed to promote interaction between both the exhibit and the viewer, and between viewers themselves. We know by watching our audience and evaluating its response to our completed exhibitions that exhibits that bring other senses into play, that create interaction or hands-on experiences, and allow public participation keep our viewer involved longer with the exhibit. I would like to see exhibits developed and designed to capture both experiences, maintaining the reverence for the object while providing an educational experience that will appeal to a broader audience.

Consider creating a strata of "learning layers," an approach that has been attempted in a variety of different ways over the years. With the use of state-of-the-art equipment, such as computers and video equipment in exhibitry, we have the technology to create multiple learning layers of information in an interactive exhibit, permitting the visitor to determine how much he or she wants to know about a subject and how involved he or she wants to become with an exhibit.

Also consider a "megapod" approach to exhibitry. No, it is not a new dinosaur, but exhibits that are developed and designed in multiple levels of information. The main traffic flow of the exhibit, the thoroughfare, consists of the real objects with exhibits of first-level information and allows the visitor to view the real specimen; pods, alleys, or multiple-learning centers that are built in architecturally, serve those individuals who prefer more in-depth learning experiences and an interactive setting.

These megapods can be developed as mini-auditoriums, interactive computer stations, open storage areas, or other methods of second- or third-level learning centers. In the case of university museums, where the curator is never at a loss for words, programs may be included to augment current course work; the public, if they choose to do so, could tap into the course-related program as well. This method of exhibitry allows our visitors, regardless of who they are, to determine for themselves how in-depth their experience will be.

PROFESSIONAL STANDARDS

Finally, I would like to touch on the direction of professional training for exhibition people. It is impossible for one individual to know everything, although I must admit, I have met a few people who think they do. It is important to be as good as you can be in your own area of expertise while having a working knowledge of the other specialties that have impact on exhibit work.

In the evolution of the exhibit process, it is no longer enough to be just a good designer. During the November 1987 Executive Board meeting of the National Association for Museum Exhibition (NAME), we asked each of our board members—in an attempt to define what an exhibit designer is—to identify ten skills it took to do his or her job. Their combined responses resulted in a list of over 65 skills that fell into four categories: creative talents, technical knowledge, administrative ability, and psychological skills. The one thing we all agreed is a prerequisite for the job is having a good sense of humor.

We have developed professional standards and ethics for exhibit-related professionals this past year and have appointed an education committee that established a curriculum to use as a standard for a graduate degree program in exhibition design for future exhibit professionals entering the field. Exhibit design professionals today come from diverse, mostly art-related backgrounds and adapt their education to fit the museum environment. But then, how many curators do you know with a degree in curating? A degree in Exhibition Design not only will help to establish professional standards for the field, but it also will help the individual's credibility when dealing with scientists or other highly credentialed individuals—even though you and I both know that professionalism is not guaranteed by the degree hanging on your wall, but by the application of your knowledge and by your quest for excellence.

A continuing-education program is underway as well, in the form of exhibition-related programs, workshops, and seminars. It is not enough only to take from your profession; exhibition professionals must give back to their profession by sharing new knowledge and technology with their colleagues and by helping to create the next generation of exhibition professionals.

Running an exhibition design department in a natural history museum is the best of all worlds; because of our position we have a foot in several camps. We are considered a public-side department, but every day we deal with the academic side of the institution. We are the main bridge between the public and the content specialists of our museum. The wealth of collection material in our museum and the visitors who walk through the front entrance are equally important to me as an exhibit designer and I have learned to respect the needs of both.

I believe that the future of exhibitions in the natural history museum setting rests on us—the museum professionals. We must focus on defining the exhibition development and design process, as well as promoting the growth and use of new technology in production methods. We just secure an in-depth knowledge of learning theory and show continued concern about conservation. We must involve ourselves with the on-going evaluation of audience responses to our work and we must make a commitment to give back to our profession by sharing information and helping to create and maintain professional standards within our profession. All of these will contribute to the professionalization of exhibition people and will heighten their determination to achieve excellence in the production of exhibits.

Acknowledgments

I would like to thank Professor Leo J. Hickey, curator of Paleobotany, who, upon his appointment as director of the Yale Peabody Museum of Natural History in 1982, allowed and encouraged exhibition team development, and Professor Willard D. Hartman, director of the Peabody Museum since 1987, for continuing to support team development, and both men for their comments during the drafting of this paper. I would also like to thank Zelda Edelson, head of publications, for her keen eye, encouragement, and gentle guidance; without Zelda none of my writings would have an ending. And most especially, I would like to thank Janet Sweeting, head of public education, who in true team spirit, sick of listening to me, patiently listens to me one more time.

Natural History Museums: Directions for Growth
Paisley S. Cato and Clyde Jones, editors
Texas Tech University Press, Lubbock, 1991, iv+252 pp.

DEVELOPMENT OF CURRICULUM-ORIENTED PROGRAMS FOR NATURAL HISTORY MUSEUMS: AN EXAMPLE IN CORPUS CHRISTI

Jane E. Deisler-Seno and Judith Reader

Abstract.—An examination of visitor statistics showed that the number of school classes visiting the Corpus Christi Museum of Science and History had dropped by approximately fifty percent over a two-year span due to internal and external factors. The museum and the Corpus Christi Independent School District cooperated to develop a curriculum-oriented science program for fourth-grade students in order to increase the relevancy of museum field trips and thereby overcome many of the factors inhibiting school visits. The revised program was designed to use the tools of the teaching profession in preparing the students and their teachers in the classroom for the trip, but still to use the traditional artifact-centered approach in the museum during the visit. Essential elements of the state-mandated curriculum are taught using a hands-on approach at the museum in conjunction with supporting classroom activities and state-adopted texts. Supporting materials include recommended pages and chapters in the adopted text and workbook, a museum-produced video tape, a lesson plan based on the video tape, a post-test, and extension activities. Museum staff also conduct two to four teacher-training sessions each year for the district in order to familiarize teachers with the museum programs. Evaluation of the program by these teachers indicates that the new program is more effective than the previous one, and that teachers are more willing to overcome the difficulties inherent in scheduling field trips as they discover the relevance of a museum visit to their curriculum.

Museums have long been involved in a "crusade against ignorance" (Bloom *et al.*, 1984). Every visitor is a potential student by virtue of having walked through the doors of the building. Because of this strong tie to education, museums have had a long history of attracting school groups into their exhibit halls. However, new restrictions on non-essentials within the school districts and concerns that students learn the basics, scheduling a field trip has become ever harder for a teacher to justify. The number of students visiting museums seems to be dropping. This was certainly the case in the Corpus Christi Museum of Science and History, an accredited general museum with a strong natural history educational history.

In 1985, the Corpus Christi Museum of Science and History reviewed its educational programs for school classes and found that the number of students attending had dropped from a high of 30,392 in the 1982–1983 school year to a low of 16,551 in 1984–1985. A variety of factors seemed to be the cause of this drop. Outside factors included declining funds available in local school districts for field trips, the loss of district funds to support a museum teacher position, and the effects of a new Texas law commonly known as House Bill 72, which restricted time away from the teaching of the state-mandated curriculum. Internal factors included lack of staff time for program development, and the quality of the programs themselves.

It was determined that, although the content of the programs was accurate, the subjects bore no relationship to the needs of the classroom teacher nor those of the students as determined by the Texas essential elements (State Board of Education, Undated). In addition, the formats of the various programs were not appropriate for the targeted grade levels, nor were the objects in the museum teaching collection utilized effectively.

With the belief that a program relevant to the needs of the teachers and their students would be more attractive to the visiting population of school classes, the museum staff approached the Corpus Christi Independent School District with the possibility of cooperatively developing a new fourth-grade science program to replace the existing one funded by the district. After school board approval was granted, the program was implemented for the first time in the 1985–1986 school year. Surveys of teachers that year and the next indicate that the methods used and the materials developed improved the program, and made teachers more willing to bring their students on a museum field trip. The methods and procedures used to revise the fourth-grade science program have since been used to develop additional programs at the museum for other grade levels.

METHODS

First, the needs of the teachers were determined through surveys supplied by the Corpus Christi Museum of Science and History and distributed by the Corpus Christi Independent School District at the end of the 1984–1985 school year. Subsequent surveys (1985–1986 and 1986–1987) were given by museum staff directly to visiting teachers who were encouraged to respond through school mail. Additional responses were gathered by randomly selecting 20 teachers per school year for follow-up interviews at their schools during their science planning periods. Museum staff also worked directly with the appropriate members of the district administrative staff, including elementary curriculum development personnel, field-trip coordinators and transportation staff, in determining program content and logistics.

Museum staff attended the district training session for the newly adopted Silver-Burdett science text books (Mallinson *et al.*, 1985) to become familiar with how teachers were to use the books and accompanying worksheets. The museum library obtained copies of these materials for reference use.

A five-minute video tape was produced by the museum with input from the district curriculum consultants. It was distributed to the schools by the district one week before a scheduled field trip. Input from surveys, teacher interviews, and the district video specialist was used ultimately in revising the content of the tape and extending it to 12 minutes.

A new bussing system was devised to shuttle one class per hour to the museum, to reduce the number of students per staff member. The new system also decreased the amount of time that busses sat idle and reduced the number of busses used, thereby lowering the cost of transportation. The museum agreed to begin the program at 9:00 A.M., one hour before normal public hours to comply with required school lunch hours.

Previsit advisory materials and postvisit testing worksheets were developed jointly by district and museum staff. The previsit materials were distributed by the museum through the interdistrict mail system and the postvisit materials were given directly to the teachers at the museum during their field trip.

Previsit materials included information about the trip and what roles the teacher, museum staff, and students were to play, a list of textbook chapters and worksheet pages to be used for introduction or review, a map of the museum, an explanation of the shuttle system, a lesson plan incorporating the video tape, an outline of the program, and supporting activities.

Postvisit materials included a worksheet to be used as a post-test, an answer sheet, a survey addressing current concerns of the district and the museum about the program, an additional sheet of optional related activities for use in the classroom to extend the lesson, and a summary of the museum presentation, including examples used, concepts addressed, and vocabulary words highlighted.

Museum staff presented two to four training classes for teachers as a part of the school district's in-service training at the beginning of each year to familiarize teachers with the fourth-grade science program. These sessions also served to acquaint them with the capabilities of the museum. In addition, the museum staff offered one to two docent-training classes per year to introduce new volunteers to the program and to refresh those currently presenting the material. Docents were observed periodically during their presentations and the observations were included in the training sessions as a means of quality control.

RESULTS AND DISCUSSION

Previsit Information

A welcome result of the new program was the strong positive reaction of the teachers. On the survey taken during the last year of the old program (the 1984–1985 school year), 79 percent of the teachers said they had no information at all about the field trip prior to the trip, and 84 percent of them wanted previsit information. After the first year of the new program (the 1985–1986 school year), only 15 percent of the teachers said they had no information about the trip beforehand, and of those who did receive the information, all found it to be very helpful and used it in conjunction with the trip. During teacher interviews, it was discovered that occasionally previsit material went astray in the office of the school itself, so distribution was changed by addressing envelopes containing the information to the "fourth grade science team leader," thereby transferring distribution responsibility directly to the teachers involved.

Two other pieces of information emerged from the interviews and the surveys. One was that teachers especially appreciated the listing of chapters and worksheet pages appropriate for preparation for the trip. Teachers also indicated that they were very glad that the museum had instituted the policy of not allowing the students into the gift shop unless the teacher had given prior permission. In previous years, teachers felt that students wasted their time in the museum buying souvenirs instead of viewing exhibits.

By the 1986–1987 school year, all of the teachers received the information, and the first, five-minute video tape had been sent out to the schools. Only 77 percent of the teachers found the previsit material to be helpful during this year, but it appears that problems with the video tape were the cause of this. Most teachers responded by survey that the video program was too short. In interviews, teachers confirmed the feeling that this was so. They asked that more scenes of the museum be added, and that the program be covered in more detail. Since then the video program was expanded to 12 minutes in length, and all of the vocabulary words to be stressed during the museum visit were superimposed on the screen. Other teachers had difficulty with the video equipment. This information was shared with the district personnel, and since then the district has updated all video equipment and added new VCRs to its inventory.

One other reason that the tape did not have the hoped-for reception by the teachers emerged: someone in the office of at least one school returned the tape before the teachers could view it. As a result, principals now are sent copies of all previsit material, including information on the tape, so that school office personnel will be more likely to expect the arrival of the tape.

Postvisit Information

During the first year of the study, no postvisit material was available. That year, 95 percent of the teachers who responded to that first survey wanted such material, and 21 percent of the respondents said that a future trip to the museum would be most improved by the addition of postvisit material. The first year that postvisit materials were available (1985–1986), 82 percent of the teachers wanted even more material.

The next year an extension activities sheet was added to the postvisit packet. This contained a variety of projects for the students to do in the classroom or as outside assignments. These activities stressed various parts of the material covered by the museum program. This time, 91 percent of the teachers used the material, including both the worksheet post-tests and the extension activities. In interviews, teachers indicated that even if they used only the worksheets, they were planning to use the extension activities in the future.

Program Content

During the last year of the old program, the students were led to an auditorium in groups of 60 to 70 and seated in chairs arranged in rows. A museum staff member would lecture the students on the importance of water to animals and man for approximately 20 minutes. A few objects were used by the lecturer, but the students had no chance to touch them. Afterwards the students were given approximately 40 minutes of free time, during which they could roam the museum at will.

Naturally, many students in the back of the auditorium did not listen to the program, either because they could not hear the speaker well, or because they were distracted by numerous animal heads on the walls, which were closer to them than the objects used by the speaker. Few of the students had the chance to interact directly with the staff member because of the size of the group and the shortness of the program. Nonetheless, 84 percent of the teachers felt that the program was adequately structured, and 89 percent felt it was aimed at the right level for their students.

During the free time, many students spent the majority of their time at the gift shop, waiting to purchase souvenirs. Others went quickly through the exhibit halls and complained of being bored. Without direction, few of the museum displays made much impression on the students.

The new program implemented in 1985–1986 was very different in content. At the museum the students were given a program that took place in three different areas of the museum, allowing staff and docents to guide student-viewing of those areas. In addition, the movement from area to area every 15 minutes helped the students maintain concentration during each segment. The students' free time was limited to 15 minutes. If the teacher gave prior approval, the students were still allowed to visit the gift shop during the free period. However, they were encouraged to go back and look at interesting exhibits they passed by during their program, something that many students did.

The program content was determined jointly by the museum and district staff. Using the adopted text as a basis, the museum staff evaluated the museum's available resources, including both appropriate exhibits and objects in the teaching collection. The district staff supplied a list of subjects from the life sciences curriculum that they felt teachers had difficulty explaining in the classroom without concrete examples.

A segment on classification was chosen because few teachers have access to the number of animal mounts available at the museum and because museums, in caring for their collections, have become specialists at classifying objects. The segment was designed around five animal mounts, four native to South Texas (white-tailed deer, bobcat, white pelican, armadillo) and one exotic species similar in appearance to cattle (cape buffalo). The students identified the familiar animals and were given some information on them and on the buffalo by the docent. Then they divided the animals and themselves into groups by what they eat. The students labeled the groups using their vocabulary words, herbivore, carnivore, and omnivore, and then reclassified these groups using two more vocabulary words, predator and prey.

The students classified all of the mounts by whether they were mammals or birds. The docent told the students what characteristics were typical of each group and encouraged the students to make their own observations. They were also asked to handle a bird bone and a mammal bone of comparable size and make observations on the differences between the two.

A second segment was on adaptation. The text covered adaptation in plants, but did not cover animals. The museum has a large diorama in the Hall of Natural History, which shows local shorebirds, including a number with specific adaptations, such as the roseate spoonbill (straining beak, wading legs and feet), the long-billed curlew (probing beak, walking legs and feet), and the double-crested cormorant (fish-catching beak, swimming feet). This exhibit was used to stress that animals also are adapted to their environments. The display also served as an introduction to the students of some of the common animals they are likely to see in the local environment. It was also used to show students how to read labels in a museum if there is no guide to answer their questions. Again, the students learned to make their own observations and compare them to familiar things, for example webbed bird feet to the flippers that people sometimes wear to swim better.

The third segment of the program covered the difference between food chains and food webs. The Hall of Marine

Science includes a large display of a number of common Gulf of Mexico game fish, a painting of a marine trophic pyramid (since replaced by a painting of a South Texas marine environment with food chain and food web interpretive labels derived from this program), and a diorama of coyotes and ground squirrels on a typical Texas beach. Here, students were able first to make a marine food chain, beginning with phytoplankton shown in the trophic pyramid and leading up through various marine species (zooplankton, shrimp, lane snapper, dorado, great barracuda, and mako shark). They practiced their classification skills by dividing the living things in their food chain into producers and consumers, thereby using two more required vocabulary words.

Next the students were directed by the docent in constructing a food web including the coyotes in the diorama. The students listed the foods the coyotes might eat at the beach, including ground squirrels, sea oat seeds, left-over picnic foods, injured shorebirds, fish washed ashore, and ghost crabs. The students then were led into realizing that many of the living things eaten by coyotes eat other things eaten by coyotes, for example, ground squirrels eat sea oats and some picnic scraps, shore birds often scavenge on animals washed ashore, as do ghost crabs, and so on, thereby creating a food web.

After a brief review and an opportunity for questions, the students were released for their free time, and were encouraged to visit the history sections of the museum, the hands-on area, and the live animals. Docents stayed in the public areas to help guide the students informally in their viewing of the exhibits, until it was time for the students to return to their school.

It was important that the program stay on schedule because the same bus that brought the students served as a shuttle. Instead of sitting idle at the museum during the hour that the program took, the bus driver went to pick up a second group of students for their program at the museum, delivering them just in time for the first group to re-board the bus and depart.

With the change to the shuttle system in 1985–1986, only one class of 25 to 30 students was in the museum at any one time, instead of three or four. The addition of docents and the division of the program into three rotating segments in different areas allowed each class to be divided into two groups, so that each docent could present the program to only 12 to 15 students at one time. This allowed the amount of staff-student contact to be increased. Students were less readily distracted from the material by virtue of this contact. To lessen distraction further in an unfamiliar environment, students were allowed a brief period at each new area to look around before the docent asked for their attention.

The end result of these changes in the program content was that on the surveys 100 percent of the teachers now responded that the program was appropriately structured and that the essential elements of the curriculum were more than adequately covered. Typical comments were that the program was excellent, "the best ever in 23 years," and that no improvement was needed.

In addition, teachers began to ask for programs in other areas, most frequently requesting programs on Indians, local history, marine science, and geology. This information has been used to develop a series of "package programs." These programs are led by the teacher, which frees the museum staff for program development. These package programs contain everything a teacher would need to lead the program, including a three to four day lesson plan with ties to the required texts, the use of the museum classroom, an appropriate object collection (or "discovery box"), guided activities in the museum, and independent activities (post-tests) for use back at the school.

The new programming appears to be achieving the original goals. Surveys and interviews show that teachers are still satisfied with the program content and format, and that they are looking for additional programs. Their view of the museum is highly positive. Visitor statistics indicate that the number of school classes visiting the museum is increasing and that more teachers are taking advantage of the museum's artifact-centered programs. The lowest number of classes visiting the museum was 16,551 in 1984–1985, with 8,585 students (52 percent) receiving programs. In 1987–1988, with all changes in place and with package programs offered for the first time, overall attendance increased to 18,205, with 10,785 students (65 percent) receiving programs. School attendance increased again in 1988-1989 to 25,643 students overall, with 15,964 receiving programs (62 percent).

Conclusion

Although a museum is largely defined by its collections, its place in the community is determined by how the museum utilizes those collections. Whether its staff is involved with the latest research or in developing new exhibits and programs, a museum is basically an educational institution. It teaches with the objects that it has, in scholarly studies of those objects published in professional journals, in exhibits developed for the typical visitor, or in hands-on programs designed for specific audiences.

Nevertheless, it is often difficult to convince classroom teachers that a museum can be an important and valuable source of practical training, in addition to the theoretical training available in the classroom. The type of learning best done in a museum is practical and hands-on, using objects more than words. Even with a museum staff member or a docent present guiding the learning process, it tends to be self-directed to a large extent. Students learn in a museum by making observations of their own. Museum staff then encourage the students to express these observations and thereby teach themselves (Pitman-Gelles, 1981). In essence, students are learning how to think, a process that is not always directly and immediately quantifiable.

In addition to restrictions placed on the teacher by the requirements of the curriculum, it is a logistical problem for the teacher to take his or her class on a field trip. Therefore, the teacher must be convinced of the value of the experience for the students to be willing to surmount the difficulties of transportation, permission slips, and so on. This can be done using the tools of the teaching profession, such as lesson plans and post-tests, which make the teacher comfortable with the educational content of the trip. This is not to say that the museum should become a part of the school district. Rather, the museum needs to communicate with the teachers, just as it communicates with any group of visitors to bring them into the building. The museum can then provide what it does best: practical, hands-on, artifact-centered learning (Heine 1976, 1977).

Additionally, many states have enacted legislation such as Texas's House Bill 72 that is designed to increase the amount of time spent by students in learning the basics. A side effect of much of this legislation is that teachers and principals are hesitant to schedule field trips because the trip may be wasteful in terms of lost learning time. If a field trip program can be developed that ensures actual learning is taking place during the trip, and that includes a way for teachers to test that learning, field trips become feasible once again.

A final difficulty is the actual cost of a field trip. Many districts have budget restrictions. However, if the school board is convinced of the educational value of the trip, and if the museum is willing to be flexible, the district may find the funds. The museum may have to make some concessions, such as opening at an unaccustomed time, blocking off an area for the class, obtaining full or partial grant funding, or absorbing some costs that the district cannot afford. However, convincing the local schools to become involved in visiting the museum is an excellent investment. The students that visit the museum on a field trip are more likely to visit again on their own, throughout their lives, thus increasing visitor support, and helping to kill the stereotype of a museum being a dusty community attic.

Acknowledgments

Many people were involved in developing and conducting this program, and it is impossible to name them all. We apologize to anyone whose name is left out. We especially would like to thank Rick Stryker, director of the Corpus Christi Museum of Science and History, the members of the Corpus Christi Independent School District Board of Education, and Betty Snow and Ceatrice Kitchen and all the other members of the C.C.I.S.D. staff involved in developing and scheduling the field trips. We would also like to thank the Corpus Christi Museum of Science and History staff and volunteers involved in or affected by the program: Lillian Bass, museum educator, for taking over and improving things, Julie Shamoun, education clerk, for holding the line when no one else was there to do it, David Rivera, foreman (and all his staff, who turn on the lights, set up the room, and put up with students in the museum too early), the docents: Mary Ellen Reinaker, Edna Hunker, Trina Grantham and Mary Anne Casstevens, and all the Gift Shop volunteers. We would also like to thank Gulf Coast Productions, for their work on the expanded video tape, Jerry Gunter, for his input on the new tape, and all the bus drivers, who put up with our shuttle system.

Literature Cited

Bloom, Joel N., E. A. Powell III, E. C. Hicks, and M. E. Munley. 1984. Museums for a new century: a report of the Commission on Museums for a New Century. Amer. Assoc. Museums, Washington, D.C., 144 pp.

Heine, A. 1976. Museums and the student. Occas. Papers Corpus Christi Mus., 2:1-18.

———. 1977. Museums and the teacher. Occas. Papers Corpus Christi Mus.,
    3:1-11.
Mallinson, G. G., J. B. Mallinson, W. L. Smallwood, C. Valentino. 1985. Silver-Burdett
    science: teacher's edition 4, centennial edition. Silver-Burdett Company,
    Morristown, New Jersey. 428 pp.
Pitman-Gelles, B. 1981. Museums, magic and children: youth education in
    museums. Assoc. Science-Technology Centers, Washington, D.C. 263 pp.
State Board of Education. Undated. State Board of Education rules for
    curriculum—principles, standards and procedures for accreditation of
    school districts. Texas Education Agency, Austin, Texas. 246 pp.

Natural History Museums: Directions for Growth
Paisley S. Cato and Clyde Jones, editors
Texas Tech University Press, Lubbock, 1991, iv+252 pp.

NATURAL HISTORY LOAN MATERIALS
FOR THE CLASSROOM

Elizabeth Patton

Abstract.—Loan materials developed by museums for use in school classrooms are becoming increasingly popular. Natural history museums can capitalize on this interest by providing accurate, attractive materials that make science fun and help reach new audiences. In planning their loan programs, natural history museums need to make decisions that will determine the style, scope, and construction of their loan kits. Recommendations are given on selecting a subject and writing the text, as well as how to choose, present, and maintain objects. Options in design and packing are examined, along with publicity, evaluation, and costs.

RATIONALE

In a 1988 survey by the Mountain-Plains Museum Education Committee, 50 percent of the respondents answered that they offered traveling kits of loan materials. Clearly, museums recognize the need for such materials and are beginning to respond. The Museum of Natural History, University of Kansas (KU), produced its first traveling kit, Mammals of Kansas, over ten years ago and has expanded the program to 15 kits on ten different topics (Patton, 1985). Although developing kits is expensive and time consuming, our experience has shown us that they are worthwhile. In 1990, we received 214 bookings that reached over 22,000 students statewide.

Why should natural history museums provide loan materials for schools? There are at least four reasons: (1) schools are often too far from a natural history museum to visit; (2) many schools may lack quality science materials; (3) field trips are costly, and schools may have to limit the number they take; and (4) museum materials in the classroom permit a relaxed and intimate experience with natural objects. This last point is important because many teachers and students are afraid of science (Bailey, 1988; Strickland, 1988), and they may be intimidated by museums (Sakofs, 1985). Offering them fun, attractive science materials in the familiar school setting can break down many barriers.

INITIAL DECISIONS

What do museums need to consider before embarking on this type of project? The first planning decision is choosing a topic. Not all natural history subjects are equally alluring to the public. A plant kit probably will not receive the positive response that a kit of mammal skins will. Listen to the prospective audience. What topics do museum visitors respond to best during the museum's summer and school year programming? Strangely enough, "creepy crawly" subjects such as spiders and snakes are very popular at KU, and loan materials offer a unique opportunity to overcome squeamishness. The subject must be not only popular, but also one for which objects are available. A dinosaur kit would be tremendously popular, but finding specimens to put in it would pose a real problem!

The next major decision involves choosing specimens. We believe that objects, be they skins, bones, or mounts, must be hands-on, and that they should be the real item, not a facsimile. Although art and history museums may be able to obtain good reproductions of artifacts, natural history objects do not lend themselves to this process as readily. Fur is difficult to fake; nothing looks or feels quite like a real feather. Choose objects that can be replaced if broken or lost. Have spares on hand, if possible. (Imagine trying to find live specimens in the middle of winter to replace a broken mount!) Some choices of objects will be constrained by size, weight, and fragility; other choices may be limited by expense. Even when prepared in-house, mammal and bird specimens are extremely costly. You have to decide how many specimens are enough, and your budget may decide this for you. Our mammal kit has 19 skins. We opted for a large number because it permits teachers to assign each student his or her own mammal, a personal and, we hope, memorable experience.

Obtaining specimens is a third consideration as you plan your loan program. Most museums have a large number of specimens that are neither research nor exhibit quaility. Most curators are glad to donate such surplus articles that have been taking up valuable space and gathering dust in their collections. Biological supply companies, such as Carolina Biological, are another source of specimens. Do not overlook the simple and the obvious: those sea shells you collected on your vacation or the seed pods you discovered on a walk through the woods can be as compelling to children as more costly items (Bloch, 1969). Road-killed animals are seldom scarce, and they can be stored in a freezer until needed.

Likewise with birds that fly into windows. Taxidermists, tanneries, and fur-trading companies may also sell skins, although the specimens may not have tails or feet unless you specifically request them.

The fourth planning decision concerns packaging individual specimens so they can withstand handling by students, as well as shipping. Bird study skins can be placed in clear plastic tubes; insects can be embedded in plastic or placed in Riker mounts; bones, after treatment with a consolidant (Storch, 1983), can be put in foam-lined boxes. Small fossils travel well in boxes with many padded compartments, much like jewelry boxes (Bloch, 1969). Even spider webs can be collected on paper and laminated (Biel, 1985). If a specimen container doesn't exist to suit our needs, we make one. Our red-tailed hawk was much larger and a different shape from any plastic tube commercially available, so one was constructed from a sheet of acetate, with end caps made from styrofoam. Nasco, a supply company in Wisconsin, provides a broad range of vertebrate and invertebrate specimens. Their Nascoguard® process results in specimens that are odorless and nontoxic; they can be stored and shipped in zip-lock plastic bags.

It is critical that museums have the appropriate permits for collecting and shipping the specimens they loan (Berger and Phillips, 1977, 1980). After seeing museum kits, teachers may want to start similar collections, and they need to be informed that collecting and licensing laws govern them as well. In addition to legal restrictions, there are some ethical considerations to be aware of. For example, some birds such as owls have spiritual significance for many groups of Native Americans, and an owl in a museum kit could seriously offend.

Remember that specimens treated with arsenic or other chemicals may present a health hazard for the students handling them. Curators may donate specimens without data and it is important to determine that they are safe for your audience (Hawks and Williams, 1986). Kits may be fumigated by sealing the container for two weeks with part of a pest strip enclosed. Air them for a week or more before sending them out (Hawks, personal communication).

TEXT

Kits generally will need some text to present information about the specimens. Teachers are afraid of being stranded with science materials they do not have the background to present, so it is important that the text be clear, concise, and easy to use. We do not give teachers a curriculum to use because that would restrict the kit to certain grade levels. Rather, we provide pertinent scientific information that can be adapted to the grade level and abilities of the students. Teachers frequently utilize this information to cover other areas, such as creative writing, reading, and spelling.

We have found a few text topics particularly important: handling instructions—"Do not remove the birds from the plastic tubes."; why this subject is important—"Why study skulls?"; a biography of the people who wrote the text, drew the posters, designed the interior, and so on. This gives the students role models and an appreciation for the work that went into the kit. A section entitled "How did they die?" should honestly and sensitively address how the specimens were collected. With potentially controversial topics, such as evolution, we neither emphasize nor omit them.

The text is written by staff in the Office of Public Education and rigorously checked for accuracy by curators and graduate students. Writing the text for natural history materials is challenging. The author must communicate technical scientific information in a nontechnical, even fun, way. Not everyone has this talent, so choose your authors carefully. If you select someone outside your museum to write a text, have them sign a contract specifying duties, completion time, and compensation. This will prevent misunderstandings and ensure that deadlines are met.

OTHER MATERIALS TO INCLUDE

Kits frequently include more than specimens and text. We want our kits to provide the teacher with materials for a week-long unit on that subject, so we include items for display, objects to touch, suggestions for activities, and books for further research. We have found that *Zoobooks* by the San Diego Zoo are an excellent resource, as are *NatureScope* activity books by the National Wildlife Federation. Filmstrips, slide sets, and posters to include are all available through commercial suppliers, such as Educational Images, the Society for Visual Education, and National Geographic, as well as state wildlife agencies. Because many study specimens do not look life-like (for example, mammal study skins are flat; preserved specimens lose color), large pictures are very important.

DESIGN

The interior design and packing plan for natural history materials can be challenging. Strange sizes and shapes seem to be the norm rather than the exception. The goal is to provide a design that protects the objects and is attractive, as well as easy to repack (Fig. 1). Soft foam layers between specimens, or compartments sunk into foam blocks work well to keep objects from shifting (Horne, 1985; Keck, 1970). Numbering and color coding help teachers use and repack the objects. In the "Mammals of Kansas" kit we drew an outline of each specimen on the fabric-covered foam it was to rest on.

If kits are to be only hand carried, the demands on the container and its contents will be much less than if shipping is involved. Cardboard may work; generally wood is much too heavy. The molded plastic containers we have purchased from Regal Plastics have proved light and nearly indestructible, both for shipping around the state by United Parcel Service and for hand carrying. One further note about shipping: United Parcel Service seems reluctant to transport dead things. Describing kits as "educational materials" seems to satisfy them if they inquire.

MAINTENANCE

Our largest maintenance problem, strangely enough, concerns not the objects, the skins, bones, and the like, but rather the paper materials that require repair and replacement. Laminating the text on sturdy cards helps extend its life. However, the folders the posters go in and the folders used for the text and slides wear out regularly even though we cover them with contact paper to keep them clean and strengthen them. Plastic folders tear and split almost as readily, especially under our extreme heat and cold. Yes, individual specimens do wear out, but in general they last for years. Careful use, along with an occasional dab of glue or a stitch with a needle and thread, greatly extends the life of the materials.

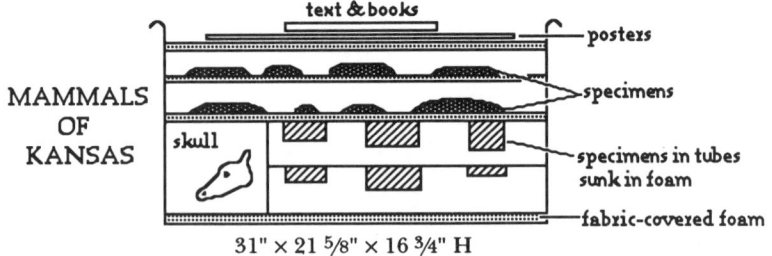

MAMMALS OF KANSAS

31" × 21 5/8" × 16 3/4" H

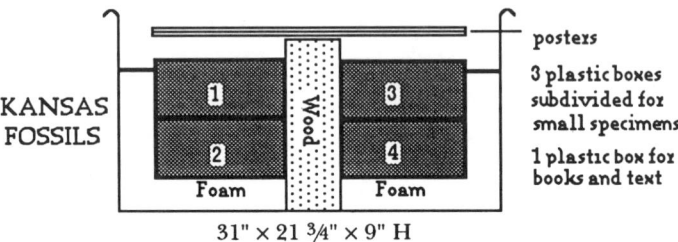

KANSAS FOSSILS

31" × 21 3/4" × 9" H

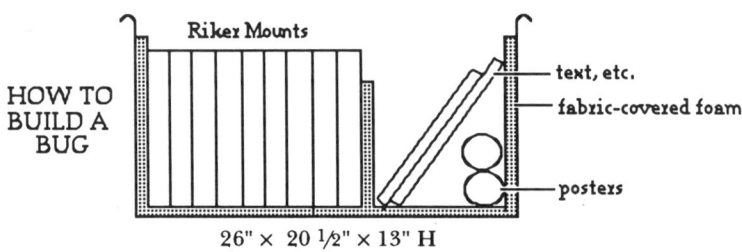

HOW TO BUILD A BUG

26" × 20 1/2" × 13" H

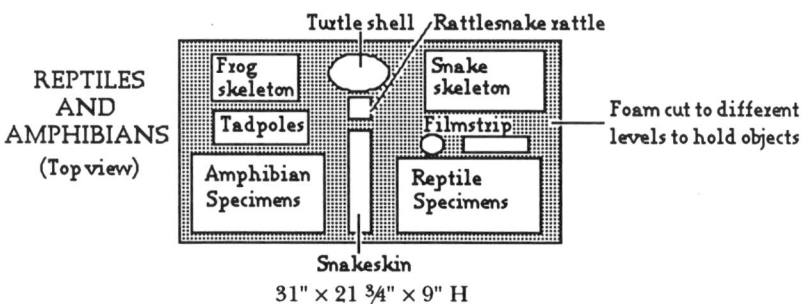

REPTILES AND AMPHIBIANS (Top view)

31" × 21 3/4" × 9" H

Fig. 1.—Packing designs.

PUBLICITY

A mailing list of schools in the immediate area is useful for publicizing your loan program, but if the option of shipping your kits exists, you need contacts around the entire state or further. The cost of sending brochures to schools statewide may be prohibitive, so you might wish to choose a subset of teachers, such as special education teachers or school librarians. For example, Kansas has two groups of science teachers, KABT (Kansas Association of Biology Teachers) and KATS (Kansas Association of Teachers of Science) that publish newsletters and readily accept museum information.

Nothing works quite as well for publicity, however, as a staffed presentation and display, such as at teacher inservice meetings. Once teachers see the actual materials and are able to ask questions, as opposed to reading a printed listing, they are eager to borrow them.

Every few years we also hold a workshop for teachers on how to use our traveling kits. "How to Borrow a Skunk: Classroom Resources from the KU Museum of Natural History" draws teachers from across the state who come to share their ideas and gain hands-on experience. These people, along with other satisfied teachers, are one of our best sources of good publicity.

EVALUATION

A brief questionnaire in the kit will provide valuable information (Fig. 2). It will tell you what you are doing right, as well as ways you can improve. It will note items that need repair ("The right leg of the prairie dog is loose."), and may contain booking requests ("Please reserve this for me next year at this time."). Be sure to ask for the total number of students that used the kit, as well as how many were gifted, how many handicapped, and so forth. This will be useful when you write your annual report for your director or look for sponsors.

COSTS

Loan kits can be very costly, the most costly aspect being, of course, labor. Kits that contain bird or mammal specimens can easily cost in excess of $3000; others, such as fish or fossils may run around $2000. You can save money if your scientific divisions donate specimens, and if you are willing to innovate

"OH, NO.!! NOT ANOTHER QUESTIONNAIRE.!!! "

WE KNOW YOU'RE BUSY, BUT WE NEED YOUR HELP.......

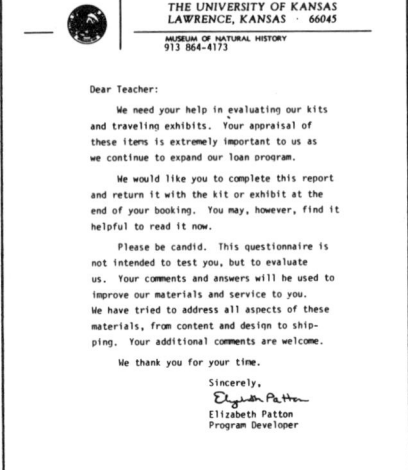

THE UNIVERSITY OF KANSAS
LAWRENCE, KANSAS · 66045
MUSEUM OF NATURAL HISTORY
913 864-4173

Dear Teacher:

We need your help in evaluating our kits and traveling exhibits. Your appraisal of these items is extremely important to us as we continue to expand our loan program.

We would like you to complete this report and return it with the kit or exhibit at the end of your booking. You may, however, find it helpful to read it now.

Please be candid. This questionnaire is not intended to test you, but to evaluate us. Your comments and answers will be used to improve our materials and service to you. We have tried to address all aspects of these materials, from content and design to shipping. Your additional comments are welcome.

We thank you for your time.

Sincerely,

Elizabeth Patton
Program Developer

I. Kit/Exhibit Descriptions
    A. How did you hear about this kit/exhibit? Check boxes below.
        1. Contact with Museum of Natural History staff □
        2. Museum publications (brochure, etc.) □
        3. Someone in your community □
           Who?
        4. Newspaper or other media □
        5. Other_____ □
    B. When you contacted the Museum did you get a sufficient and
        accurate description of the kit/exhibit? □ Yes   □ No
    C. Were the expenses what you expected based on the description?
        □ Yes   □ No
    D. Do you have any suggestions for ways to publicize these
        materials?
II. Content and Design
    A. How did you use this kit/exhibit? Check boxes below.
        1. Display □
        2. Individual study □
        3. Group activities □
        4. Other_____ □
    B. Did you use the text? □ Yes   □ No
    C. Was the text:
        1. Easy to use □ Yes □ No
        2. Informative □ Yes □ No
        3. Appropriate length □ Yes □ No
    D. Check the areas you would like to see more fully covered:
        1. More objects □
        2. More text □
        3. More activities □
        4. Other_____ □
    E. Did you develop your own materials? □ Yes   □ No
        If so, describe_____
    F. Suggestions

                                        continued...

III. Shipping
     A. Did you receive enough information about the following:
         1. Shipping □ Yes □ No
         2. Packing □ Yes □ No
         3. Arrival and due dates □ Yes □ No
     B. Did you receive the kit/exhibit on or before the date promised?
         □ Yes □ No
     C. Did the kit/exhibit arrive in good condition? □ Yes   □ No
     D. Suggestions
IV. Audience
     A. Estimate how many people saw or used the kit/exhibit_____
     B. What grade levels?_____
     C. Did more than one school use the kit/exhibit? □ Yes   □ No
         If so, how many?_____
     D. Did more than one teacher use the kit/exhibit? □ Yes   □ No
         If so, how many?_____
     E. Were these materials used by the gifted/talented? □ Yes □ No
                                        the handicapped? □ Yes □ No
     F. Would you recommend this kit/exhibit to others in your
         community?
     G. Do you plan to reserve this kit/exhibit in the future? □ Yes □ No

Comments and Suggestions:

Your Name:
Name of Kit or Exhibit:

FIG 2.—Kit questionnaire.

and use materials in unusual ways (Bloch, 1969). For example, an ordinary electric carving knife can be used to shape large blocks of foam. The life span of a well-constructed kit is, fortunately, quite long. Our oldest and most popular kit, Mammals of Kansas, is 13 years old and still in use.

We charge a small user fee ($25 for one week) to cover maintenance and shipping. Although charging more would help defray initial development costs, we are concerned that it might also put our kits beyond the reach of the audience that needs them most.

CONCLUSION

Natural history loan materials provide an important educational function for schools that cannot visit museums, yet want an enjoyable, economical, and high quality science experience for their students. Museums wishing to provide loan kits to schools face a number of initial planning decisions: choosing a topic, selecting specimens, and obtaining and packaging the specimens. They then need to consider the kind of text to include, as well as ancillary materials such as slides, books, or tapes. Design and packing of the shipping container, along with maintenance, should be carefully planned to ensure a long and useful life for the kit. Publicity, evaluation, and costs also need to be considered so that the museum reaches its audience and remains responsive to it. Museums should investigate successful loan programs at other museums to help them with their planning and implementation.

LITERATURE CITED

Bailey, A. L. 1988. Ending science illiteracy. Museum News, 66(5):50-53.

Berger, T. J. 1980. Index to U.S. federal wildlife regulations: an annotated reference. Association of Systematic Collections, Lawrence, Kansas, 400+ pp.

Berger, T. and J. Phillips. 1977. Index to U.S. federal wildlife regulations. Association of Systematics Collections, Lawrence, Kansas, 270 pp.

Biel, T. 1985. Spiders. Zoobooks 2. Wildlife Education Ltd., San Diego, 18 pp.

Bloch, L. 1969. The red box. Museum News, 47(7):29-31.

Hawks, C. and S. Williams. 1986. Arsenic in natural history collections. Leather Conservation News 2(2):1-4.

Horne, S. 1985. Way to go! Crating artwork for travel. Gallery Association of New York State, Hamilton, New York, 53 pp.

Keck, C. 1970. Safeguard your collection in travel. American Association for State and Local History, Nashville, 78 pp.

Patton, Elizabeth. 1985. The museum goes to school: creating loan materials for
    teachers. Museum News, 63(5):17-22.
Sakofs, M. 1985. Structuring a museum in-service for teachers. J. Mus. Ed.:
    Roundtable Rep., 10(2):16-17.
Strickland, C. 1988. Science for Kids: Winning Formulas. Foundation News,
    29(5):20-25.
Storch, P. 1983. Field and laboratory methods for handling osseous materials.
    Conservation Notes, No. 6.

APPENDIX

Suppliers

Tube-Paks (tubing and end caps)
Fantastik Plastics
3124 Gillham Plaza
Kansas City, MO 64109
(816) 561-0402

Shipping Containers
Regal Plastics
4405 E. 11th
Kansas City, MO 64127
(816) 483-3040
(800) 852-1556

Specimens
Carolina Biological Supply Company
Burlington, NC 27215
(919) 584-0381
(800) 334-5551

Nasco
901 Janesville Ave.
Ft. Atkinson, WI 53538
(414) 563-2446
(800) 558-9595

Publications and Other Visual Aids

Zoobooks
Wildlife Education, Ltd.
930 W. Washington St.
San Diego, CA 92103

National Geographic
Education Services
Washington, D.C. 20036
(800) 368-2728

NatureScope
National Wildlife Federation
1412 16th St. NW
Washington, D.C. 20036

Society for Visual Education
1345 Diversey Parkway
Chicago, IL 60614
(312) 525-1500
(800) 621-1900

Educational Images
P.O. Box 3456, West Side
Elmira, NY 14905
(607) 732-1090
(800) 527-4264

Natural History Museums: Directions for Growth
Paisley S. Cato and Clyde Jones, editors
Texas Tech University Press, Lubbock, 1991, iv+252 pp.

TRAVELING EXHIBITS AS A STRATEGY FOR UNIVERSITY-STATE MUSEUMS OF NATURAL HISTORY

Peter B. Tirrell

Abstract.—Traveling exhibits appear to be a successful strategy for raising the profile of and increasing public support for a university-state museum of natural history, especially in the mountain-plains or midwestern states that have rural populations and few large metropolitan areas. The Oklahoma Museum of Natural History has developed a system that has reached over two million people in 14 states at 370 sites. Sites have included schools, libraries, and museums in small communities, as well as in metropolitan areas. Exhibit projects have generated over $750,000 in grants and earned income. The system provides extensive public service throughout the state. The grant funds and creative activities have attracted the attention of the museum's parent organization, the University of Oklahoma, and legislators in the state government. The museum believes that the exhibits have been especially helpful in achieving its new status as the state museum of natural history and in reaching its primary goal of obtaining a new facility.

This paper presents a case study regarding the planning process and results of the traveling exhibit program that has been designed and developed by the Oklahoma Museum of Natural History, Norman, Oklahoma. The museum was formerly known as the Stovall Museum of Science and History, an organized research unit of the University of Oklahoma. It is the museum's belief that the traveling exhibits have been an important strategy in reaching its goals and meeting its needs. These include obtaining a new status as the state museum of natural history, improving its financial support base from the university, and seeking funds for a new facility.

Historical background.—In 1978, the museum had an opportunity to join the Oklahoma Archaeological Survey and two Native American groups in the planning and development of two traveling exhibits. Each exhibit focused on the historical and present-day culture of two Native American tribes, the Plains Apache and the Wichita. Planning funds from the National Endowment for the Humanities (NEH) already had been received. The museum was asked to lend its educational and exhibit expertise to strengthen the proposal for implementation funds of nearly $200,000. The museum agreed to participate in the project. Grant funding was received from NEH and the project was successfully carried out. Two separate exhibits were produced, one representing each tribe. The exhibits opened at the museum in the summer of

1982 followed by a year-long tour at 14 other sites in Oklahoma. With additional funding from NEH, the exhibits were combined and displayed at sites in four other states. At the end of the regional tour, the exhibits became part of the permanent display at the Apache and Wichita tribal centers, respectively. By 1984, the public and the university reactions to the traveling exhibit were so positive that a traveling exhibit system was proposed as part of the museum's long-range plan. The museum selected traveling exhibits as a strategy after reviewing its mission, goals, needs, and a variety of external and internal factors that influenced the museum's activities.

TRAVELING EXHIBITS AS A DEVELOPMENT STRATEGY

The museum's decision to develop a traveling exhibit system was carried out through strategic planning. The pros and cons of traveling exhibits were carefully considered. An environmental assessment had to be made, that is, a determination of the internal and external factors that governed or influenced the museum's directions and activities. The museum also needed a perspective on its physical and institutional characteristics, its needs and goals, and its financial outlook. In 1978, when the museum began to consider traveling exhibits, many university museums, including Stovall, had financial pressures that resulted in cutbacks in facilities, services, and staff (Davis, 1976). Colleges and universities suffered severly at this time, and in turn their museums faced difficult decisions. Campus museums of art, history, and natural science in many states, including Texas, Louisiana, Washington, Michigan, Idaho, and New Jersey, were either closed or had their programs drastically curtailed (Black, 1984). Professional museum consultants indicated that the Stovall museum was in desperate need of a new building, more staff and curators, and a 100 percent budget increase (Black and Tordoff, 1980). The Stovall museum had just enough funding to keep its telephones ringing, its staff working, its exhibit galleries open, and its leaky roofs repaired, while trying to protect five million objects in the collections. Thus, the opportunity to get project funds via a traveling exhibit was especially tempting. Projects such as those carried out with the Apache and Wichita tribes looked like a "gift horse" not to be looked in the mouth. Why did we look? The answer is that financial planning is not museum planning, and budgets are not plans nor strategies for achieving a museum's purpose. The best strategic plan is one that defines the comprehensive focus and thrust of a museum's actions and resources to create its most advantageous position in the future (McHugh, 1980). The museum had to ascertain that the development of a traveling exhibit program was going to help its long-range plans, not just get a grant or make a few more shows. The long-range needs were (1) a new museum facility, (2) an improvement of its status in the parent organization, (3) recognition as the state museum of natural history, (4) more support, especially for maintenance and operations, (5) a broader base of support, including state funding, (6) an increase in the number of staff and curators, and (7) a comprehensive public service program.

The pros and cons of traveling exhibits.—In its planning, the museum had to determine the pros and cons of traveling exhibits before it could assess their potential in the context of the museum's needs and characteristics. Several publications were especially helpful in this process, including the *Manual of Traveling Exhibitions* (Osborn, 1953). Additional information was found in Grobman (1958), Hudson (1966), Keck (1970), Shaeffer (1971), Bergmann (1976), and Wolf (1976). The Museum also contacted various museums and organizations that had developed traveling exhibits, including the Smithsonian Institution Traveling Exhibit Service (SITES). The following list of general pros and cons was made from the source information. The museum also listed pros and cons that seemed to be unique to its status as a university museum.

Pros—these will happen for the museum:
  1. increase visibility in community,
  2. increase audience,
  3. increase use of objects,
  4. increase breadth of program,
  5. increase in exhibit space.

Pros—these may happen, unique to each museum:
  6. reach special interest groups or minorities,
  7. grant funding,
  8. cooperative efforts with other agencies,
  9. increase professional standing,
 10. increase support,
 11. fewer cuts in staff, funding.

Cons—these will happen for museum:
1. objects are subject to increased damage or loss,
2. increase in complexity of program, recordkeeping, communication,
3. increase in insurance,
4. increase in management,
5. scheduling and shipping,
6. increase in maintenance.

Cons—these may happen, unique to each museum:
7. decrease in professional standing,
8. management, leadership, and personnel problems,
9. need for addtional skills, knowledge,
10. conflict with mission.

Identifying and assessing external and internal factors.—In selecting a strategic plan, the museum had to identify and assess internal and external factors. External factors included areas such as the museum's audience, public image, relationship to other museums, geographic position, and demographic characteristics, including those of special interest groups, minorities and metropolitan and rural populations. Internal factors included areas such as the museum's mission, resources, facilities, and governance. Many of the external and internal factors were similar to those of other university-state museums of natural history in the mountain-plains or midwestern regions. As discussed by McHugh (1980), it became clear that the museum had to define its special mission and establish a comprehensive focus and thrust for its actions and resources. These tasks have not been easy for state-university museums. Williams (1969) and Nicholson (1971) indicated that university museums usually are forced to serve two masters, whether they want to or not. The university's priorities focus on students, teaching, research, publications, and obtaining extramural grants. The state's priorities include access to objects, answers to all questions of natural history, recreation, and attracting tourism dollars into the state. Whether the university museum is officially the state museum of natural history is irrelevant. The academicians and their students teach, research, and publish, primarily for the benefit of other academicians and their students. However, school-age children may want to know about dinosaurs, adults may look for family entertainment, and special interest groups such as Native American legislators may ask what the museum is doing to share its collections and expertise with local tribes and nations. The role of the university museum has not always been clear to administrators, faculty, or even to museum staffs and curators themselves (Freedman-Harvey, 1989). Coleman (1942) expressed that the university museum's principal and overriding charge was to serve the university community, its students, and faculty. University museums that deny this role lose the support of faculty and administrators (Black, 1984). Coleman also indicated that the public service functions could be more important depending upon local factors, such as the number and accessibility of other museums in the area. However, serving the university does not mean that the goals and purpose of the museum are, or should be, the same as those of relevant teaching departments of divisions of the university (Black, 1984). Collections of objects, artifacts, artworks, and specimens are fundamental to a museum's mission and focus. Collections may or may not be fundamental to a departmental agenda. A natural history museum may have a strong, integral relationship with a department of zoology that emphasizes systematic and field studies of vertebrates. However, it is not likely to have a strong working relationship with a department that emphasizes cellular and molecular studies. Thus, the University may consider the museum a cornerstone of its teaching and scholarly research or a millstone of antiquated practices and curios. The public may consider it a front-line battle station or the door to an ivory tower (Williams, 1969).

In 1984, the following external and internal factors appeared to be most important to the museum's future:

A. External
  1. Demographics.—The museum existed in a largely rural state with only four defined Standard Metropolitan Statistical Areas (SMSA) (Bureau of the Census, U.S. Department of Commerce, 1980). A large population of Native Americans (lived) in the state, only half of whom lived in an SMSA near the museum Thus, a traveling exhibits program would provide museum outreach to rural and Native American Communities in Oklahoma.
  2. Special characteristics and significance.—The museum existed as virtually the only museum of natural history in Oklahoma, with the largest and most comprehensive

collection of any type. The public demanded to see
more of the museum's collections, to paraphrase one
complaint: "You came and dug it up, took it away, and
nobody ever gets to see it anymore!" The Museum had
the opportunity to meet the public's strong interest in
the museum's resources and interpretive services.

3. Relationship with university departments and special
research units.—The museum had a good working
relationship with several university departments, research units, and state offices, including the Department of Zoology, the Oklahoma Archeological Survey,
and the State Archeologist. The museum could anticipate
cooperation and a sharing of resources from them.

4. Public image.—The image of the museum was most
frequently associated with old dinosaur bones in old
buildings where everything stayed the same. The
public also viewed the museum as a service agency of
the state because it was part of the university, a state-
supported institution. The museum needed to make
people aware that it had a great deal to offer, especially with a change of status.

B. Internal

1. Mission.—In its simplest form in 1984, to collect,
preserve, research, and interpret the natural and cultural history of Oklahoma and the region.

2. Parent institution-governance.—Interests of the
University of Oklahoma included teaching, research,
publications, and grant funding.

3. Collections.—There were over five million objects in
the collection valued at more than $100 million.
These include the 15th largest collection of vertebrate fossils in the United States, a nationally
famous collection of prehistoric Native American artifacts from the state, and the most extensive collections of mammals, birds, and herptiles in the state.

4. Staff and curators.—There was a professional staff
that had the skills to design, construct, and manage
the system, but additional academic and support personnel were needed.

5. Budget.—Approximately 85 percent of the museum's
support came from the university.

6. Facilities.—The museum had 45,000 sq. ft. of space, approximately 20 percent of what was needed. Housing was in eight separate buildings, none designed as museum structures. Several buildings were temporary structures built for use by Navy operations during WWII. Burn-down time for these buildings was eight minutes. Less than 4500 sq. ft. (10 percnet) was available for exhibits. This exhibit space was very small when compared with similar institutions (Nevling and Smith, 1987).

Rationale for selecting traveling exhibits as a strategy.—As discussed, the museum wanted a strategy that would create an advantageous position in the future. The traveling exhibit strategy was selected for the following basic reasons: (1) the exhibits were the best strategy based on the museum's internal and external environment; (2) the opportunity for funding was favorable; (3) the exhibits would produce tangible results that could be quantified to indicate the museum's direction; (4) the strategy seemed to offer the most concrete, visible results in the academic as well as the public community, and (5) there seemed to be a special opportunity for cooperative efforts and a sharing of resources with special interst groups, academic departments, and governmental agencies or offices. Several other alternatives were considered, including areas of membership, research, publication, and educational programs and on-site exhibits. None of these appeared to have the potential of traveling exhibits.

Specific objectives of the system.—A strategy tells you what you are going to accomplish and how (McHugh, 1980). The museum made the following list of objectives to accomplish after assessing the pros and cons of a traveling exhibit within the needs and unique characteristics of its time and environment.

1. To reach schools, libraries, and museums in small communities, as well as large metropolitan areas;
2. to have artifact-oriented exhibits that emphasized the strength of the collections;
3. to produce the exhibits with grant funds, gifts, donations, or earned income;
4. to have cooperative efforts with other institutions and organizations;
5. to design the exhibits to be modular and reusable;

6. to reach minority groups, including Native Americans and rural communities;
7. to be self-reliant, and carry out the program from proposal to completion;
8. to generate funds from the exhibits as seed funds for more exhibits;
9. to have the exhibits evaluated;
10. to make the exhibits highly visible;
11. to have educational and audiovisual materials accompany the exhibits;
12. to keep careful records of exhibit sites and attendance;
13. to present the exhibits as creative activities.

RESULTS

In the past eight years, the museum has developed 11 exhibits that have reached over two million people in 370 sites (showings) in 14 states, including 233 sites in Oklahoma. Currently, there are four exhibits on tour. The exhibits have ranged in size from approximately 500 sq. ft. to 2000 sq. ft. All but two art exhibits, have been object-oriented, primarily high-quality replicas to reduce security needs. Interpretive labels, photographs, graphics, hands-on materials, audiovisuals, and teachers' guides accompany the exhibits. The format is free-standing panels and cases. Specially designed crates are used for shipping. Scheduling and shipping are either carried out by the museum staff or the TRACKS system of the Oklahoma Museums Association. The museum has obtained three trucks for shipping exhibits, partly through exhibit grants or income generated through the exhibit system.

The average annual attendance at host sites represents an increase of 150 percent in visitation to museum exhibits. Sites have included small, rural communities, as well as metropolitan centers, including 103 museums, 94 schools, 71 libraries, and 102 others comprised of galleries, tribal centers, banks, community centers, and colleges and universities. The exhibits were successful in reaching Native Americans in local Oklahoma communities such as Apache, Anadarko, Carnegie, and Okmulgee. By request, the exhibits and accompanying educational activities were presented at national meetings of Native Americans such as the Native American Community Health and Recreation Conference, San Antonio, Texas, and the Indian Education Conference, Tulsa, Oklahoma.

By placing exhibits at other sites, the museum's effective exhibit space has been increased approximately 111 percent from 4500 sq. ft. to 9500 sq. ft. The system provides extensive public service throughout the state, reaching 120 communities and all but 14 of 77 counties (Fig. 1). Outside of Oklahoma, the exhibits have had a regional distribution with several distant sites (Fig. 2). The projects have generated over $750,000 from 12 grants and earned income including rental fees for exhibits. Funding agencies have included the National Endowment for the Humanities (NEH), the Oklahoma Foundation for the Humanities (OFH), the Kerr Foundation, the Noble Foundation, and AT&T. Several exhibits were jointly sponsored by the Oklahoma Archaeological Survey, the Plains Apache Tribe, the Wichita Tribe, and the Muscogee-Creek Nation. Two of the exhibit designs received Special Merit Awards from the OFH.

The grants, earned income, and creative activities attracted the attention of the university administration. Due to its high rate of grant success, the museum received lower budget cuts than other offices and departments. In addition, travel funds were made available to the staff for presentations associated with the exhibits. The staff has made nine presentations or workshops at professional meetings related to the exhibits.

In the early stages of development, design and management problems arose. Changes in administrative and exhibit personnel took place. Permanent staff were often overextended when support personnel were lost as grant funds ended. The new administration and personnel established the museum's leadership in carrying out the program. Use of the TRACKS systems of scheduling and shipping has eased problems in these areas where support personnel were important. High-quality replicas were used for most exhibits instead of original items. Little damage occurred with the objects or the exhibit materials. No negative evaluations have been received from the sites, visitors, funding agencies, or professional evaluators.

In regard to its long-range plan, the museum became the state museum of natural history for Oklahoma as of 1 July 1987, by way of a landslide vote in the state legislature. In addition, the University of Oklahoma Board of Regents has approved a site and architectural master plan for a new museum complex that will have 300,000 sq. ft. and cost $40 million. A

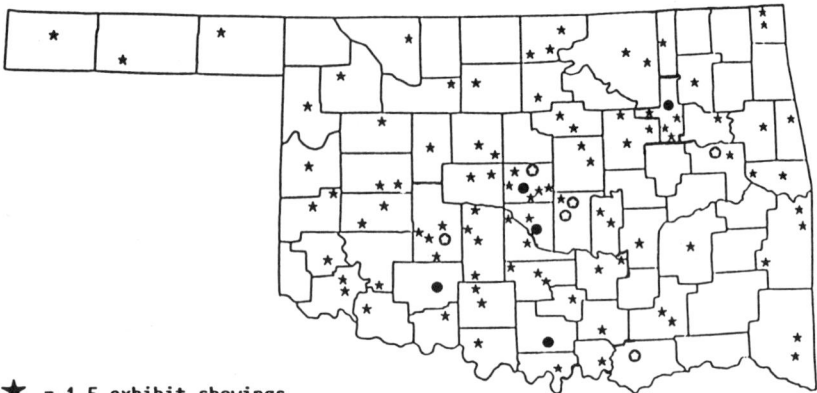

★ = 1-5 exhibit showings

✪ = 5-10 exhibit showings

● = 10 or more exhibit showings

FIG. 1.—Distribution of traveling exhibits of the Oklahoma Museum of Natural History in Oklahoma, 1982–1990.

FIG. 2.—Distribution of traveling exhibits of the Oklahoma Museum of Natural History in the United States excluding Oklahoma, 1982–1990.

board for the development of this complex is forming. Building funds will be sought from a combination of private, state, university, and federal sources. The university marked the museum as a priority in its Centennial Funding Plan. In 1988, the museum was also reaccredited by the American Association of Museums. Several new curatorial and staff positions have also been added with permanent university funding.

DISCUSSION

The traveling exhibit system has been a successful strategy in helping to reach the long-term needs and goals of the museum. The direct effect has included results such as increased funding and attendance. The exhibits have also kept the museum in a favorable environment and position, especially in the minds of university administrators and state legislators. Through the exhibits, the museum has satisfied the university's focus on grants and creative activities. By placing exhibits in the hands of the public, the museum has satisfied the need for access, information, and recreation, especially in areas where there is no other museum or people are unable to visit the Oklahoma Museum of Natural History. In contrast, the exhibits have also tended to overshadow other public programs, and this has been negatively criticized by reviewers of proposals for General Operating Support from the Institute of Museum Services. Additional new public programs are being added to provide overall balance. However, the museum's long range-plans include the development of one additional traveling exhibit per year. Several are in stages of planning, development, design, and construction. The museum's traveling exhibit system should serve as a model that is useful to other university-state museums of natural history. As indicated by Conrad (1990), university museums have the potential to be much more innovative than other museums. The model may be most useful to the museums in the mountain-plains and midwestern regions that share similar demographics and characteristics as the Oklahoma Museum of Natural History. In addition, the system is adaptable to use by a consortium of such museums in these regions, an interest of the Oklahoma Museum of Natural History.

ACKNOWLEDGMENTS

I thank Michael A. Mares for editorial comments and suggestions, and the following for their valuable assistance in planning and carrying out the exhibit program: Judy Furlong, Judy Jordan, Mike McCarty, Carolyn Pool, Roger Vandiver, Dan Timmons, and Rick Whitehead. A special thanks is given to participating institutions and agencies including the Apache Tribe of Oklahoma, the Wichita and Affiliated Tribes, the Muscogee-Creek Nation, and the Oklahoma Archaeological Survey. A personal thanks to Pat Smith, Jane Tague, and Pam Rector for their assistance and patience in preparing this paper for publication. Funding for the exhibits was obtained in part from the AT&T Foundation, the Oklahoma Foundation for the Humanities, the National Endowment for the Humanities, the Kerr Foundation, the Noble Foundation, and the Norman Arts and Humanities Council. Additional contributions were received from the W. H. Brine Company, STX-INC., and the Frank Paxton Lumber Company.

LITERATURE CITED

Bergmann, E. 1976. Exhibits: a proposal for guidelines. Curator, 19:151-156.

Black, C. C. 1984. Dilemma for campus museums: open door or ivory tower? Mus. Stud. J. I(4):20-23.

Black, C. C., and M. B. Tordoff. 1980. Evaluation report: Stovall Museum. Unpubl. 6 pp.

Davis, G. 1976. Financial problems facing college and university museums. Curator, 19:116-122.

Freedman-Harvey, G. 1989. University museums and accreditation. ACUMG Newsletter VI(1):5-7.

Grobman, A. B. 1958. Museum extension through traveling exhibits. Curator, 4:82-88.

Hudson, D. S. 1966. The artmobile: fine art on the road. Curator, 9:337-355.

Keck, C. K. 1970. Safeguarding your collection in travel. Am. Assoc. State and Local Hist., Nashville, Tenn., 78 pp.

McHugh, A. 1980. Strategic planning for museums. Mus. News, 58(6):23-29.

Nevling, L. I., and C. B. Smith. 1987. A report prepared for the Museum Accreditation Program, American Association of Museums. Unpubl. 25 pp.

Nicholson, T. D. 1971. A question of function. Curator, 14:7-10.

Osborn, E. C. 1953. Manual of traveling exhibitions. Imprimeric Union, Paris, 112 pp.

Shaeffer, M. W. M. 1971. Not all Indians live in tipis: local prehistory sparks a museum education program. Curator, 14:184-193.

United States Department of Commerce, Bureau of the Census. 1980. 1980 census of the population, general social and economic characteristics, Oklahoma. U.S. Government Printing Office, Washington, D.C., Vol. 1., C-38, 489 pp.

Williams, S. 1969. A university museum today. Curator 12:293-306.

Wolf, S. J. 1976. Guidelines for traveling exhibits. Curator, 19:226-232.

Natural History Museums: Directions for Growth
Paisley S. Cato and Clyde Jones, editors
Texas Tech University Press, Lubbock, 1991, iv+252 pp.

THE ROLE OF NATURAL HISTORY MUSEUMS IN IMPROVING SCIENCE EDUCATION IN RURAL SCHOOLS

Jeffry Gottfried, Rebecca Smith, and Judy Dacus

Abstract.—The New Mexico Rural Science Education Project (NMRSEP) serves as a model for museum outreach programming. Under National Science Foundation sponsorship, project staff from the New Mexico Museum of Natural History and the Center For Rural Education at New Mexico State University have traveled tens of thousands of miles to instruct rural teachers in how to teach science using local natural resources. The NMRSEP approach capitalizes on the natural settings of rural schools and the unique capabilities of museums of natural history. Through participation in workshops, field trips, and natural history field schools (outdoor natural science festivals), rural teachers learn about the wildlife, plants, rocks, and fossils of their area and how to develop science activities centering on locally available natural history specimens and other inexpensive items.

Although much has been written bemoaning the problems inherent in teaching science in small and rural schools, natural history museums can turn rural school liabilities into science education opportunities. Because of school size and other factors related to their rural situation, a lack of funds for laboratories and equipment and other science facilities are often cited as major stumbling blocks in the quest for quality science programs in rural schools. The problems that exist and the discrepancies in funding and school facilities between larger urban and smaller rural schools seem overwhelming. However, to natural history museum educators, charged with conducting outreach programs to rural communities, the outlook for rural science education is brighter. In fact, rural schools have many distinct advantages over their urban counterparts.

1. Rural schools are located near natural areas that provide the raw materials for outdoor laboratories in the study of zoology, botany, geology, soil science, hydrology, and ecology. No amount of public funding or specialized equipment can provide these facilities and resources to urban schools.

2. Rural children grow up playing in natural environments and start their formal schooling with much implicit knowledge and familiarity with nature and natural processes.

3. Smaller rural schools often have the increased flexibility in
   scheduling needed to put into motion the logistics necessary
   for conducting outdoor science studies. The task of the
   rural science educator, in our view, is to capitalize on these
   unique advantages to the betterment of their programs.

Notice that we refer to the natural environment as raw
material for outdoor laboratories. This is because there is a
giant leap from having the natural resources close at hand to
having the knowledge and skill needed to create a curriculum
that effectively utilizes them to teach the skills, processes, and
concepts of science. What is needed on the part of rural
science educators is a very specific knowledge of the plants,
animals, rocks, minerals, soils, weather, and ecology of their
communities. It is the use of rock from the school yard or
children's backyards, soil or water samples from the family
farm, and insects and wildlife from the local community
(pointing out the special significance of the mundane) as opposed to generic specimens purchased from a catalog, which
excites students to study and learn about the natural sciences.
When science learning is tied to local animals, plants, or
geologic formations, students take their learning home with
them and share it with their families. Their daily travels
about their community reinforce their school learning.

Typically, college courses and the text books used in the
natural sciences first deal with subject matter generalities,
citing examples from Africa, Asia, and the Galapagos Islands
and not the local school community. How can teachers obtain
the information that they need to bring science to life for
their students using local resources in science teaching? Our
experience tells us that natural history museums can play an
important informational and instructional role.

THE NEW MEXICO RURAL SCIENCE EDUCATION PROJECT
A Model for Improving Science Education in Rural Schools

Natural history museums possess the knowledge, skills, and
personnel needed by rural schools in order to utilize their
abundant natural resources in teaching science to children.
They are staffed with curators who are experts in the natural
sciences, especially as they pertain to local environments, and
educators whose job it is to interpret the natural sciences to
the public. In the state of New Mexico, the New Mexico

Museum of Natural History has joined forces with the Center For Rural Education at New Mexico State University, experts in the field of rural education, to create the New Mexico Rural Science Education Project (NMRSEP). Developed by a grant from the National Science Foundation (MDR 8550535), NMRSEP's goal is to improve science education in rural schools. Since June, 1986, a team of science educators, scientists, and rural education specialists have traveled tens of thousands of miles to provide science education training to rural teachers and students in five rural schools working toward the following goals.

1. To make teachers aware of their locally available natural resources.
2. To provide teachers with specific information about their local natural resources (wildlife, plants, geology, paleontology, weather phenomena, biotic communities).
3. To provide teachers with locally collected specimens useful in teaching science.
4. To correlate existing science goals and textbook chapters with locally available natural resources.
5. To help teachers to develop specific activities and experiments that require only inexpensive materials, and that utilize their local environment in achieving their science teaching goals.
6. To achieve one through five above in such a way as to make the program replicable in other states, thus creating a model for rural science education.

Museum Resources

The NMRSEP outreach activities are carried out using resources commonly found in natural history museums. In this regard, the NMRSEP serves as a replicable model for other museums.

Personnel.—A two-year National Science Foundation grant ($356,859) provided for a research staff of six plus a secretary, from June, 1986 to June, 1988. NMRSEP staff included a core of natural history educators from the New Mexico Museum of Natural History and specialists in rural education and curriculum development and an evaluator from the New Mexico Center For Rural Education at New Mexico State University. Museum curators of paleontology, geology, zoology, and botany were also involved on a consulting basis. A secretary handled scheduling, correspondence, and word processing.

Vehicles.—A passenger van, covered pick-up truck, and four-wheel drive station wagon were used in various combinations for transporting staff, participating teachers, specimens, and equipment.

Equipment.—An Apple-Macintosh SE and printer were used in the preparation of educational materials and correspondence. Equipment used in field activities included binoculars, rock hammers, live traps, plant presses, nets (insect and aquatic), microscopes, hand lenses, aquaria, and maps. A video camera, monitor, 35 mm (OM-2s) camera, and slide projector were routinely used in the project.

Natural history collections.—Hands-on specimens used in outreach efforts were those from the museum's Education Department Collection, which consists of over 1000 cataloged geological, paleontological, botanical, and zoological specimens and an herbarium with over 300 specimens. Specimens in the education collection were deemed unsuitable for inclusion in the permanent collection of the museum by the curators.

New Mexico's Unique Natural History Resources

For the reader to appreciate fully the essentially untapped science resources available in the state of New Mexico, we would like to share the following facts about New Mexico's natural history: (1) the state of New Mexico has tremendous geographic diversity. Its lowest elevation is 2500 ft.; its highest is 13,500 ft. (2) It is comprised of five natural regions: (a) the Colorado Plateau, (b) the Great Plains, (c) the Rocky Mountains, (d) the Basin and Range Province, and (e) the Mogollon Highlands. (3) Seven ecological zones occur in the state (alpine, spruce-fir, mixed conifer, ponderosa forest, woodlands, grassland, desert). In many areas of the state, it is relatively easy to travel from desert to alpine forest and even tundra in a matter of a few hours. (4) Rock outcroppings in the state document two billion years of natural history and earth processes, including the extinction of the dinosaurs and transition to a mammal-dominated landscape. Clearly, from this brief list, there is a tremendous amount of raw material for science study in New Mexico. What about your own state? It may not have the diversity of New Mexico but certainly it possesses many unique and fascinating natural resources that

TABLE 1.—*Demographic data for participating schools shown in Figure 1.*

School District	Number Enrolled in Grades 1-6	Student Ethnicity			Teacher Ethnicity		
		% Anglo	% Hispanic	% Native American	% Anglo	% Hispanic	% Native American
Cuba	196	9.7	34.9	55.4	31.6	57.9	10.5
Dulce	259	0.6	9.3	90.1	37.8	54.1	5.4
Estancia	322	46.5	52.8	-	75.6	11.1	-
Hatch	492	22.6	77.4	-	71.6	28.4	-
Magdalena	151	20.9	35.5	43.6	64.3	35.7	-

complement the science goals mandated by your state, exemplify the concepts that teachers need to teach, and are inherently interesting to your state's students. As a state or local natural history museum, you can be of help in getting teachers started.

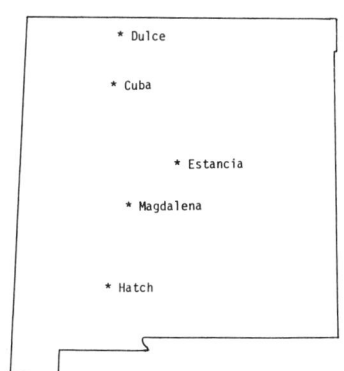

FIG. 1.—Locations of the five participating schools in the first two years of NMRSEP.

Our Study Sites.—We began our project by inviting five school districts to participate in the project. Districts were selected so as to represent New Mexico's geographic, geologic, ecological, and cultural diversity. Figure 1 shows the locations of the participating schools, and Table 1 shows their demographic data.

Year-One Activities

In the early months of the NMRSEP, project staff developed and distributed sample kits narrowly focusing upon local natural history themes to be used as models by the elementary teachers. Day-long workshops were scheduled with teachers at each of the participating schools. During the workshops, project personnel introduced information about local natural history and exhibited sample kits based upon local environments. A workshop was conducted by a project staff member from the Center For Rural Education on teaching strategies and curriculum development. Staff members also gave seminars on how to develop a science kit using local natural resources and on expected outcomes of the project. A resource inventory of each participating school's community was conducted starting with the area near the school. The surveys and background materials (for example, local geological maps) were gleaned from a wide variety of technical publications and other sources, rewritten by project staff for the layman, and compiled into natural history booklets. Teachers also participated in these surveys that documented the natural history of their communities, the materials available, and their potential for teaching science to children. Bird and mammal sightings, tracks, the occurrence of plants, and geologic setting were noted and presented to the teachers as the raw materials for kits. At the end of the workshops, regular meetings were scheduled with the teachers during which teachers and project staff reviewed the progress of development of the kits. Typically, project staff left these meetings with an assignment from teachers to collect information on some relevant topic. In this way, project staff and participating teachers functioned as curriculum development teams.

Teachers were expected to develop kits over a period of approximately one year. During this time, field trips to local sites were conducted for teachers led by experts in specific areas of natural history. Although one teacher might request a specific field trip pertaining to the development of his or her kit, typically his or her colleagues, their families and students participated as well. In addition, a special summer field trip program lasting one to two days was offered during the summer of 1987. In this way, teachers in each district participated in a series of field trips, each aimed at a number of different natural resources of their community.

Teachers were given information on developing a natural history teaching kit from *Your Community as a Natural Science Laboratory* (Gottfried and Smith, 1988), to assist in planning their own kits. Two training sessions on teaching science process skills were presented to teachers by project staff from the New Mexico Center For Rural Education so that they could develop science activities that taught the science process skills mandated by the New Mexico State Department of Education. A kit contains a collection of natural history specimens, some background material on the subject, a set of classroom activities based on some aspect of the local natural history, and resources for students wishing to learn more about the specific topic. In the process of teaching about specific natural science concepts, kits also teach science process skills such as observation, experimentation, hypothesis formulation, and communication.

Topics for the natural history kits that were completed by rural teachers and NMRSEP staff, and now being utilized in the schools are:

1. Trees.—Teachers in three school districts chose to develop kits on trees, which included collections of foliage, bark, cones, photos, field guides, cross-sections of trees, and a set of activities focusing on local species.

2. Cacti.—There is an abundance of cacti species along the nature trail adjacent to the school in Hatch where this kit, which includes information and activities focusing on the biology and ecology of cacti, was developed.

3. Wildflowers.—This kit includes activities that focus on the parts of a flower and on flower adaptations of local plants, examples of pressed specimens of wildflowers, and instruction on how to build a plant press and start a classroom herbarium.

4. Seeds.—This kit includes a collection of seeds from locally occurring plants and activities on seed anatomy, adaptation, dispersal mechanisms, and germination.

5. Ants.—This kit utilizes the most ubiquitous natural resource available to teachers in New Mexico—ants! It includes background information and instructions for conducting studies of ant anatomy, solar architecture in ant hills, ant foraging and feeding behavior, and orientation in ants.

6. Aquatic Life.—This kit, which focuses attention on the animals and plants that inhabit a spring-fed pond in an otherwise dry area, includes activities on aquatic insects, crayfish, algae, and aquatic microorganisms and information on setting up a pond-life aquarium for classroom observation.

7. Mammals.—This kit includes a small collection of mammal skulls and stuffed live-mounts prepared at the museum, background material on mammals of New Mexico divided according to plant communities, and activities on skull and teeth adaptations, mammal tracks, temperature regulation, and habitat preference.

8. Spiders.—This kit includes a small collection of preserved spiders and activities on web-building behavior, feeding behavior, and locomotion.

9. Birds.—This kit gives instructions for setting up bird-feeding stations, and lists the commonly occurring species in the area, and includes collections of bird feathers and nests, along with activities that use these materials.

10. Weather.—This kit provides background information on meteorology, with special reference to phenomena common in the area. The kit makes use of five years of weather data for the area and challenges students to draw inferences and make predictions based upon these data.

Project personnel provided resource materials, such as videotapes, books, and articles, as requested by teachers. Staff helped with the identification of specimens collected for the kits. They also provided extensive curriculum development, editing, and word-processing services for the teachers. Each participating school was given a metal specimen cabinet, obtained by project staff from the New Mexico State surplus yard. To varying degrees, schools have added rocks, minerals, fossils, and other natural history specimens to their cabinets for use in science teaching.

Origin of the natural history field school concept.—At the end of the first year, project personnel were invited to the end-of-the-year picnic held at a National Forest campground near the Cuba Elementary School. Staff members organized a table at the picnic containing natural history specimens (live snake, fossils, rocks, bones) and equipment (microscope, insect net) and conducted several activities relating to the natural history of the immediate area, including a natural history scavenger hunt. The students' response was overwhelmingly enthusiastic, choosing to stand in line for some time to view plants and insects under a microscope, borrow an insect net, or examine fossils and rocks from the area first-hand. Teachers also expressed surprise at how captivated their students were, choosing to take part in science activities as opposed to playing in the stream, fishing, or playing softball. After all, this was the last day of school! Teachers expressed interest in having project personnel return at a later time and do more activities with their students. Ultimately, the idea of demonstrating to teachers how to teach science using local natural resources was developed more formally into natural history field schools that were scheduled for the second year of the project.

Second Year Activities

By the end of the first year, it was apparent that continuing to prod teachers into producing natural science kits useful to other teachers would be non-productive. Although teachers continued to be enthusiastic about the natural science workshops, demonstrations, and field trips conducted by project staff, they clearly did not have the time nor the writing and research skills necessary to create science kits and develop activities, even with support of project staff. In many cases, teachers' writing and punctuation were so poor as to necessitate total rewriting by project staff. In three separate instances, teachers expressed feelings of guilt and having "let down" project staff in not completing projects. What was needed was a means of instructing teachers about local natural history and science teaching methods in the way that we would like them to teach their students. The program had to take less time on the part of teachers. From the Cuba Elementary School picnic, we learned that both students and teachers enjoy learning science in outdoor settings and are very curious about local natural history. Therefore, staff systematically developed plans for natural history field schools, outdoor natural science festivals that focused on the local environment.

Field schools.—NMRSEP began its second year with the beginning of the 1987–1988 school year. Project personnel chose to begin the year with natural history field schools held for each of the participating schools. In response to the lessons learned in year one, the field schools were planned so that teachers and students participated in activities, teachers learned in an appropriate context to be carried over to the classroom, and a festival-like atmosphere was associated with pleasurable science activities. For each school, a nearby site was chosen for the field school locations, typically a National Forest or tribal campground. Project staff arrived at the site the day before the field school was to be held, surveyed the natural history of the area, set out live traps for small mammals, and collected specimens for use in the next day's activities. Project staff camped overnight at the site. Typically, all students at a rural school attended the field school for one of the two days that it was in session. Older students (fourth, fifth, sixth graders) participated on one day and younger students on the other. As students arrived at field school, they were greeted by banners and balloons and welcomed by project staff. Students were separated into groups of approximately 15 and given colorful buttons with local natural history images on them. The color-coded buttons identified the various groups and helped in maintaining order; however, they came to be treasured by students and staff as a memento of the day.

Groups of students rotated from activity station to activity station on a time schedule that gave older students 25–30 minutes for an activity and younger students 20 minutes. Teachers were divided among the groups so that they could also participate in the activities. Parents were invited to attend and helped monitor student behavior so that teachers could be active participants in field school. Each group participated in six or seven activities, depending on time constraints of the day. Activity topics included plants, mammals, insects, geology, birds, pond life, weather, fossils, owls, nature sketching, and trees. Not all activity stations were available at each field school. Rather, each field school was designed to take advantage of the particular natural resources of each community and field school site.

After field schools, packets of information and suggested follow-up activities were sent to teachers for their own personal information and for use in the classroom. Project staff each collected questions that were asked during the course of the field school and sent answers and a bibliography of suggested readings to teachers. Project staff also conducted follow-up meetings with the teachers to encourage and motivate them to use these activities and to work with the teachers to incorporate the activities within the established science curriculum for the school. On three occasions, parents communicated with project staff after field school and requested information and that they be invited to attend future sessions.

During the second year of the project's development, staff members also provided individualized assistance on natural history topics of special interest or significance to particular teachers.

The Summer Elementary Science Teachers' Institute

NMRSEP responded to teachers' requests for more in-depth instruction in natural science field activities by staging our most ambitious workshop to date, called the Summer Elementary Science Teachers' Institute (SESTI). SESTI was a five-day residential workshop from June 12–17, 1988, offered to selected teachers from each of the nine rural school districts that had chosen to participate in the NMRSEP during the 1988–89 school year. In addition to project staff from the museum and the Center For Rural Education, SESTI attracted many volunteer faculty, including the museum's curator of zoology, an education specialist at the museum who specializes in integrating art and nature, a mathematician at the University of New Mexico and amateur astronomer, a soil conservationist from the U.S.D.A. Soil Conservation Service, a wildlife biologist with the New Mexico Department of Game and Fish, and a forester with the U.S. Forest Service. All guest instructors participated at no cost to the project and with permission of their employers. At SESTI, teachers and staff participated in a wide variety of science activities and explored the natural resources of the area around the Circle A Ranch in the Nacimiento Mountains just east of Cuba, New Mexico. Activities included measuring stream flow, microhabitat studies, nature art, natural history games, insect identification, fossil collection techniques, soil sifting and classifying, skull comparisons, astronomy, and others. Through these activities, teachers gained extensive hands-on experiences in natural science field studies. It was an opportunity for project staff to offer rural teachers the best of the ideas, activities, and information developed over the course of the past two years of National Science Foundation sponsorship. It was also the end of the research and development phase of the NMRSEP under NSF sponsorship and the start of the service phase of the project under sponsorship of the New Mexico Commission on Higher Education/U.S. Office of Education.

Each activity at SESTI focused on the immediate environment of the Circle A Ranch and vicinity. However, skills and activities learned during the week were transferable to each of the participating school districts. Teachers from each district met periodically during the week in order to adapt activities and approaches to their home environments and resources.

For example, every school represented at SESTI could conduct studies of locally occurring aquatic life. Some could study streams, some ponds, and some seasonal irrigation ditches or vernal pools. The species of organisms that would be expected to be found would certainly differ, but their ecological niches and the types of student activities used to explore the environments could remain more or less the same. Following each SESTI session, teachers began to systematically adapt activities to their home environments. After SESTI, teachers were given an environmental survey form to complete that identified local natural resources and opportunities for activities and needs for adapting existing ones. The teachers' environmental surveys almost always brought forth many new questions and requests for information.

EVALUATION

First Year Activities (1986–1987)

Methodology.—The final evaluation of the first year of the project was performed using a modified transactional evaluation. Teachers in Cuba, Dulce, and Estancia were interviewed, eliciting comments about activities in which they participated. A questionnaire (see Appendix) was then developed from comments made in these two schools. The questionnaire asked respondents to indicate whether or not they strongly agree, slightly agree, slightly disagree, or strongly disagree with 20 statements about activities in the preceding year and 10 statements about activities for the coming year. The questionnaire was used in the southernmost schools (Hatch and Magdalena) participating in the project. Responses were scored from one to four with strongly agree given a score of one and strongly disagree given a score of four. Only those items with an average score of less than 2.0 or greater than 3.0 are reported here.

Table 2 presents the statements concerning first year activities that received average scores of less than 2.0 from the teachers in Hatch and Magdalena, indicating agreement among the teachers. In Hatch, six teachers who had been involved in the project for the year responded. It should be noted in interpreting the results that personnel from the Museum of Natural History were primarily involved with the schools at which teachers were interviewed, whereas the New

TABLE 2.—*Statements with which teachers in Hatch and Magdalena agree (average score; response of 1.0 indicates total agreement, 4.0 indicates highest disagreement).*

| | Average response | |
| | Hatch | Magdalena |
Statement	*(n=6)*	*(n=5)*
I enjoyed doing the research on my kit.	1.2	1.5
The support from the rural Science Education and Museum people was very good.	1.0	1.0
I would have liked a detailed outline of what was needed for putting together a kit.	1.8	1.3
I really enjoyed the trips the teachers took out in the field.	1.0	1.0
This project is helping me get an idea of what to do when I take my students outside.	1.3	1.5
I feel that working on the kits was very worthwhile.	1.2	1.3
I am more enthusiastic about science since being in this project.	1.5	1.5
I have tried to incorporate things I have done in the project into my classroom activities.	1.3	>2.0
Working with this project helped me be more ware of the features of our local environment.	1.0	>2.0
I find it hard to devote time after school to working on the kit.	1.3	>2.0
I spend more time now having my students do hands-on activities than I did before being involved in this project.	1.7	>2.0
I feel that working with the project has made me a better science teacher.	1.8	>2.0

Mexico Center for Rural Education provided personnel who were primarily involved with the schools responding to this questionnaire. Teachers in Hatch responded with agreement to many statements made by teachers in interviews at Cuba, Dulce, and Estancia. Most indicated that they have enjoyed being involved with the project. Although teachers agree that it was difficult to devote time after school to work on kits, they also agree that it was worthwhile to do this activity. Several teachers indicated in marginal notes that they were already doing hands-on activities (not necessarily related to natural history) before being involved in the project and therefore did not feel that the project had strongly affected their use of these types of activities in the classroom. Teachers in Hatch indicated disagreement with only two statements as shown by an average response of greater than 3.0. The two statements were: "Sometimes I felt there was a lack of communication between our teachers and the project personnel" (average response

3.6) and "Sometimes the project personnel mentioned doing things that they never followed up" (average response 3.7). The lesser communication problems with Hatch may be due to its accessibility to Project staff at New Mexico State University when compared with that of the the northern schools. The Hatch school was visited with more regularity and frequency by staff from New Mexico State University due to its proximity to Las Cruces than the northern schools, which are more distant from the museum.

In Magdalena, five teachers who participated in the project for the 1986–87 school year responded. Teachers in the Magdalena school did not agree with as many items as the teachers in Hatch (see Table 2). On those statements with which they did agree, however, the average response was the same or nearly the same as that from the teachers in Hatch. The teachers in Magdalena, based on their responses, were less affected by their participation in the project. Only one statement of the last six statements in Table 2 received an average response of less than 2.0. All but one of these statements concerns the effect of the project on the teachers. The apparent lack of effect of the project on Magdalena teachers may be due to their late entry into the program and an initial lack of commitment and on-going lack of enthusiasm. For example, only in Magdalena did teachers stand up and leave workshop sessions at 4:00 pm. Only in Magdalena did only three faculty members take advantage of a free two-day summer workshop and field trip. Teachers' morale and attitudes toward improving their science teaching skills seemed to be much different in Magdalena than in other schools. The fact that the Magdalena schools has had three superintendents in as many years and has been taken over in recent years by the State Department of Education for failure to meet minimum educational standards may also serve to explain teachers' lack of enthusiasm for donating their time for the purpose of improving science teaching skills.

The teachers in Magdalena indicated disagreement with only one statement as shown by an average response of greater than 3.0. The statement was, "Sometimes the project personnel mentioned doing things that they never followed up" (average response 3.2). During the meeting in which teachers responded to the questionnaire and then participated in a discussion session, it became obvious that teachers in the Magdalena school were less enthusiastic about the project than teachers in other schools. In the meetings in Cuba, Estancia, and Hatch, teachers were outspoken in their likes and dislikes about the project. During the meeting in Magdalena, it was difficult to get any response from the teachers. The school principal was in the room part of the time but his presence did not seem to make a difference either in the teachers' responses or lack of responses.

Because of the personnel involved, each school reacted differently to the project. The enthusiasm or involvement of the principal did not appear to have a major effect on the attitudes of the teachers. In some project schools, the principal appeared to be uninterested, whereas in others the principal was an active participant in workshops. There was no way to predict the attitudes of the teachers from the apparent attitudes of the principals. The apparently least interested principal had enthusiastic teachers, one very enthusiastic principal had the least enthusiastic teachers, and in other cases the interest and enthusiasm of the principal approximated the interest and enthusiasm of the teachers. Sometimes, teachers expressed enthusiasm in the presence of project staff members and then did not carry through between visits. Project personnel found this frustrating as it slowed down project activities. Eventually, though, they realized that most teachers did not have either the time or skills to take the lead in developing kits. It should be mentioned here that there were some outstanding exceptions to this rule. Three teachers at Estancia Elementary School enthusiastically produced natural science kits and developed field trip and classroom activities that were all very imaginative and effective.

Findings.—Teachers expressed both positive and negative reactions to the general atmosphere of the project, the development of natural science kits to be used in the classroom, and outdoor activities presented by project personnel. Teachers' comments indicated that they were very receptive to the overall project. They enjoyed learning about their local natural history, enjoyed working with project personnel and reported that the personnel were very helpful in providing information and feedback. They stated that project personnel were very responsive to the needs and concerns of the teachers. Project personnel were also viewed as enthusiastic about the project and subject matter which helped keep the teachers involved. The development of kits was enjoyable for some teachers who reported that they had enjoyed the research conducted in developing the kit. Teachers stated that doing the kits was worthwhile. Most expressed interest in seeing the kits that were developed at other schools. From the point of view of project staff, however, having teachers develop natural history kits was not viewed as a successful strategy for producing coherent, well-organized materials that could be shared between teachers within a district. Even with the help of project staff and the support of the museum, less than 43 percent of teachers completed a science kit as described above. It was not until year two that project staff discovered many positive outcomes of the teachers' science kits development project. During year two, long after project staff had given up on the concept of teachers developing science kits, we discovered that approximately 60 percent of participating teachers were still collecting background materials for their kits, as well as natural history specimens. What was even more important, they were teaching science units to their students and using what they had developed. None of these "long-lost" kits was anything to look at—boxes of specimens and file folders of copies of articles and activities. However, project staff was extremely impressed that teachers had persevered on their own long after the deadline for completing kits had passed. They had created materials that were apparently of use to them. Conversations with these teachers indicated that they had learned a great deal from having worked on their kits. They were very enthusiastic about them. What they had objected to was having to write and develop science activities in a polished form that others would read and use.

Other feedback on teachers' kits came from their students. On a field trip to the Manzano Mountains conducted for his students by Mr. Wieland Elstner, a participating fifth-grade teacher who had created a kit on the trees and life zones of the Manzano Mountains, students were able to explain to project staff the distribution of trees on the mountain and identify the relevant variables affecting tree distribution. They were also able to identify trees to species, explain what lichens were, and seemed quite knowlegeable about many topics covered in Elstner's kit.

Project staff visited the classroom of Donna Neish, a participating fourth-grade teacher who had created a kit on aquatic life focused on a pond within walking distance from school. Ms. Neish's students were eager to show off the pond-life aquarium that they had created using organisms that they and their teacher had collected. Students were able to point out aquatic insects, such as damselflies and caddis flies, and were able to explain complete and incomplete metamorphosis and the difference between insect larvae and nymphs. They also pointed out to us the adaptations of diving beetles for breathing underwater. It was clear that the project had been of great interest to them and was stressed by their teacher.

Students of Eula Newton (third grade) and Kathy Williams (kindergarten) demonstrated their knowledge of ants and other social insects to project staff when they visited Estancia Elementary School to install an ant colony exhibit. Students could identify the queen ant and could explain her role and the roles of other ants in the colony. Student art work and specimens that they brought to show project staff indicated that they had spent considerable time and effort studying insects, the topic of their teachers' science kit.

The natural history field trips offered were enjoyable to teachers. Project personnel and teachers spent the day immersing themselves (at times literally!) in the local environment and its natural history. Following these experiences teachers stated that they were now more aware of their local environment and what it had to offer them and their students in the way of science activities and opportunities for study.

The most frequently voiced criticism of the project was a perceived lack of focus. Staff believed this problem to be caused by a difference in perception and expectations between them and the teachers. Teachers viewed the project as primarily a service to them. Project staff, on the other hand, viewed the project as a series of experiments in teacher training in science education; they had many ideas that they wished to try out. When one approach, procedure, or technique did not appear to be fruitful, they dropped it and moved on to another. The problem of teacher satisfaction could be remedied by more advanced planning on the part of project staff and more frequent updates with teachers on the state of the project, and rationale for future plans.

A criticism voiced by one school was lack of communication between the teachers and the project personnel. Further investigation indicated that the major problem was within the school administration. Project personnel were communicating with the school principal, but the information was not always reaching the teachers. This could be remedied by having a teacher contact who received the same information as the principal.

According to statements made by teachers at the two schools, the development of the kits was hampered by a lack of direction from project personnel. Teachers stated that they would like a detailed plan of how to complete a kit. They would also find it helpful to be given firm deadlines for finishing certain tasks. Other teachers did not want deadlines because they felt it pressured them into doing something they might not have time to do. Because this was a research project and development was to be an ongoing activity, some of the project personnel chose to be flexible and not to hand teachers a formula for kit development. It is clear that many teachers would have been more comfortable with established guidelines for the development of kits. Certainly, objectives should be established for the finished product and presented to the teachers who will work on the kits. In fact, teachers participating in the first year of the project were the experimental subjects that provided guidelines, hints, and suggestions for future science-kit development by teachers.

Second Year Activities (1987–1988)

Field schools.—Teachers and parents were asked to assist with the evaluation of the field schools by responding to several questions; additional comments were invited also. Each person was given a card containing the questions as they began each station and cards from previous stations were collected at the same time. A copy of the evaluation card is presented in Table 3.

Response rate was 96 percent. Nonresponses occurred only when the teacher or parent had to leave the station to discipline a student or to escort younger ones to restrooms, neither of which happened frequently.

Table 4 indicates the percentage of responses to each question from the field schools held in Dulce, Estancia, and Magdalena.

TABLE 3.—*Evaluation card (original size— 4 x 6 cards).*

Check the appropriate space.	
1. Are you a teacher _ a parent _.	
2. Do you feel students in your group learned at this station:	
how to observe the natural environment?	Yes _ No _
how to properly collect a science specimen?	Yes _ No _
how to work as a group?	Yes _ No _
how to protect our natural environment?	Yes _ No _
that science can be fun?	Yes _ No _
3. Will you as a teacher or parent try to discuss activities presented	
at this station in your class or home?	Yes _ No _
4. Was the amount of time allowed at this station:	
too long _ about right _ too short _	
5. Considering your group, was material presented at this station:	
too complicated _ about right _ too simple _	
6. Would you like handouts and other materials on this topic to use	
in your class or home?	Yes _ No _

The Cuba field school was not evaluated in this manner. Because there were no significant differences in responses from the field schools, data from the three have been combined (two adults per group responded at each station at the three field schools—a total of 219 responses were collected).

Although evaluation cards were designed to evaluate each station as well as the overall field school, there were few differences in the data collected at each station. The questions concerning what students learned were developed from comments from the project personnel during a meeting in which the field schools and project outcomes were discussed. Not all of these were applicable to a particular activity. For example, at the birds, owls, microclimates, and trees stations, as well as others, no specimens were collected.

The low percentage of "yes" answers on protecting our natural environment is a result of several factors. In some cases, information about protecting the environment was presented subtly. In another, the staff member was heard discussing the use of live traps so the small mammals collected would not be harmed. One response card at the end of that particular session had checked "no" to the question about protecting the natural environment. Perhaps teachers and parents evaluating the session were distracted from presentations and did not hear remarks about protecting the environment or perhaps they did not connect the remarks with this

TABLE 4.—*Evaluation of field schools (sample size=219).*

Statement	Percent Responding			
	Yes	No	No	N/A
Students learned to observe the natural environment	92.8	3.0	4.8	
Students learned how to properly collect a science specimen	76.4	10.9	10.9	5.5
Students learned how to work as a group	82.9	9.1	7.3	3.0
Students learned to protect the natural environment	55.5	15.9	25.0	3.7
Students learned that science can be fun	98.2	0.6	1.2	
Will you as a teacher or parent try to discuss activities presented at this station in your class or home?	89.6	0.6	9.8	
Was the amount of time allowed at this station				
too long?	5.6			
about right?	67.7			
too short?	26.7			
Considering your group, was material presented at this station				
too complicated?	6.9			
about right?	91.9			
too simple?	1.3			
Would you like handouts and other materials on this topic to use?	92.9	0.96.3		

*Some respondents wrote that this was not applicable.

item. Hearing these types of remarks also requires a level of awareness that some teachers and parents may not have had.

Follow-up is important for children's learning. It is encouraging to see that both teachers and parents indicated that they would try to discuss activities presented at the stations. To encourage follow-up activities, project staff mailed packets of information to each school with specific answers to questions that were asked during field school, instructions on how to conduct each of the field school activities, and background information on each of the activity stations. The final question concerning handouts and other materials referred to these materials. Teachers were almost universally eager to obtain these follow-up materials.

The amount of time was considered to be about right or a little too short. There was no one station that elicited either a too-long or too-short response from all evaluators. In fact, the same station sometimes received a too-long response from one evaluator and a too-short response from another on the same day. The time schedule was set up at the two-day field schools so that older students had 30 minutes at each station plus five minutes to change from one station to another. The younger students had 20 minutes at each station plus the five minutes to change stations. At the one-day field school, all students spent 25 minutes at each station.

Activities and information were considered by most respondents to be at the appropriate level for the age of the students involved. Project personnel changed activities from day-to-day for the different grade levels attending. Occasionally, activities would be modified during the day if necessary to meet the students' needs.

In addition to checking appropriate spaces on the cards, parents and teachers wrote comments on their cards. Also, one school had participated in the field school prior to the interviews mentioned previously, and during the interview teachers made additional comments about the field schools. The most frequent comment was that there should be a day of field school just for teachers. Some suggested that the teachers should be allowed to have time to see the activities before bringing the students. Teachers suggested that if they had received more information before the field school, they could have better prepared their students to obtain the greatest information from each station. Teachers also stated that they needed to see all of the activities because it was a problem for them to know how to respond if some of their students asked questions about a station not seen by the teacher.

Teachers and parents were very complimentary about the field school, stating that they had learned new things along with the students and that they had really enjoyed the day. The enthusiasm of parents was very important. Many times project personnel could not distinguish between teachers and parents. Sometimes the parents were more involved and enthusiastic than teachers. Because parental involvement with schools plays a role in improving the schools, field schools might be used at some later date for the purpose of increasing parental involvement in school or for fostering positive parent-child-teacher interaction.

A variety of suggestions was offered for improving the field school. Teachers wanted more information about the logistics of the field school prior to arriving at the site. Both teachers and parents commented that students were distracted by other near-by groups. This was solved by moving stations further apart. All comments were constructive, offering the project personnel information on ways to make the field schools better. Many useful suggestions from field-school participants were put into effect in later field schools. Students were interviewed informally throughout the day during breaks and at lunch time. Most enjoyed the field school. Some wanted more time at stations in which they had the most interest. Discipline problems were minimal, indicating that students were participating at each station according to instructions given by the person in charge of the station.

Generally, the field schools were viewed very favorably by teachers, parents, and students. Staff found field schools to be most rewarding and most exhausting, taking many days to prepared for and many long hours to conduct for an entire school. If schools had to pay the actual cost for a field school, they would never be able to afford to have one conducted by the museum. Therefore, to make these worthwhile and educationally effective events more accessible and affordable to rural teachers, project staff wrote a chapter on how to conduct a natural history field school in the booklet, "Your Community as a Natural Science Laboratory" (Gottfried and Smith, 1988).

Natural history field schools were universally popular and well-received. Teachers would like more of these days, including some that are specifically for teachers to participate in without students being present. This suggestion led to the creation of the Summer Elementary Science Teachers' Institute.

The Summer Elementary Science Teachers' Institute.—The second year of the project ended with SESTI, a five-day science teaching institute that is described in the project narrative. This program introduced NMRSEP to new teachers, while allowing teachers who had participated in the past to further assimilate and apply what they had learned. The following is a summary of the evaluation of SESTI.

Participating teachers were given a science subject matter pre- and post-test that had been developed by SESTI instructors based upon the course syllabus. They were also queried on a number of other topics relating to the selection of topics for SESTI, the enjoyment of activities, the relative usefulness of the sessions to teachers and suggestions for improvements for next year's program.

SESTI
Participant Scores

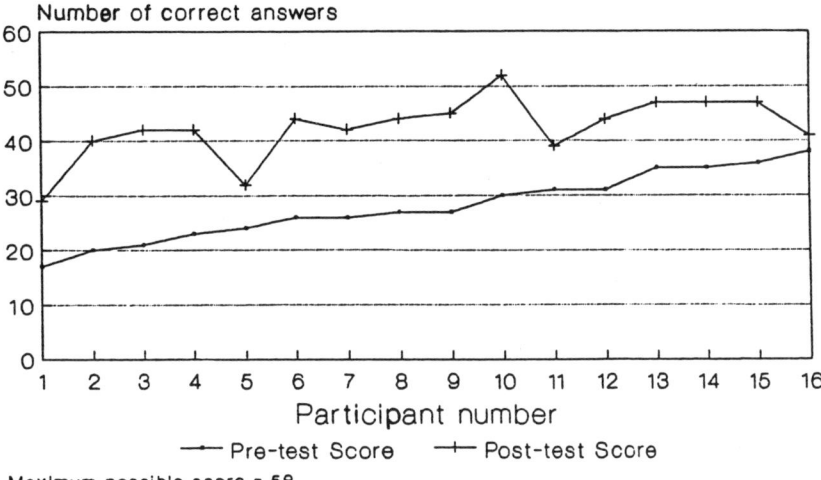

FIG. 2.—Gains made between pre- and post-test scores by SESTI participants.

Figure 2 illustrates the gains made between pre- and post-test scores. As can be seen from the graph, teachers' varied greatly in their familiarity with the subject matter when they arrived at SESTI. However, all teachers improved their scores significantly over the course of the week. Teachers also demonstrated that they learned many new things at SESTI as judged by their poems, essays, and sketches in *The Natural Inquirer,* a newspaper produced at the end of the week at SESTI.

Equally as important as learning subject matter, teachers' responses to evaluations and their open-ended comments were universally positive about the science activities and SESTI, in general. This enthusiasm hopefully will carry over to the classroom and will be translated into more and better science activities for students. It is significant to note that following SESTI, participating teachers developed and carried out their own follow-up activities and have made plans and commitments that will impact the teaching of science.

1. Teachers from Los Lunas District planned two field schools. They also began a natural resource survey.

2. Teachers from Zuni Pueblo participated in a science fair for teachers, held prior to the student science fair. Teachers selected their science fair topics as a result of their SESTI activities.
3. Linda Soloman, a participating teacher whose specialty is language arts, has mailed information on the "Whole Language Program" to other SESTI participants, along with her recommendations for integrating science and language arts.
4. Two participating teachers schedule dates for their students to exhibit natural history art at the museum's "1801 Gallery of Children's Art."
5. Four SESTI participants have contacted Rolene Barnett, editor of the *Naturalist Center News*, a publication of the New Mexico Museum of Natural History, written by children for children. They received guidelines for the publication and all of them are planning on having their students submit writings to the publication.

The various spin-offs from SESTI including such things as personal contacts and the items listed above ultimately may be the most important outcomes of SESTI. Teachers who participated in SESTI will also serve as instructors along with NMRSEP staff in workshops for their fellow teachers held in their districts. In this way, SESTI has had a multiplying effect in each school district.

Additional Outcomes of the NMRSEP

Certain significant data on the impact of this project were collected outside the formal evaluation process. These points warrant mentioning.

1. Although 46 percent of participating teachers did not complete natural science kits that were transferable or useful to other teachers, 66 percent completed kits that they used in their own classes during year two of the project. Writing materials for other teachers turned out to be a more difficult stumbling block (too time consuming for teachers) than development of "unpolished" kits for a teacher's own use.
2. Two of five schools participating in the development phase of the NMRSEP (1986–88) opted to pay a $2000

fee in order to continue their participation during the 1988–89 school year.

3. Teachers at Cuba Elementary School applied for a $3000 grant to construct a nature trail on school property. NMRSEP staff will assist Cuba teachers in developing the trail during the 1988–89 school year.

4. As a result of its participation in NMRSEP, Hatch Elementary School applied for and received a grant of $2000 from the Title II program (U.S. Office of Education), which was spent on science equipment and reference books.

5. Project staff and teachers at Estancia Elementary School designed and built an experimental ant colony to complement the natural science kit on ants developed by Estancia teachers. The ant colony is on display in the library of Estancia Elementary School. If the ant colony and associated activities proves successful at Estancia, other such colonies will be constructed in participating schools.

6. An exhibit on determining the age of an elk from its teeth was constructed for Dulce Elementary School in cooperation with the Jicarilla Game and Fish.

7. Because so many teachers chose this topic, project staff developed a traveling exhibit titled "Life Zones and Trees of New Mexico" showing rainfall, temperatures, and elevations of an idealized New Mexican mountain range. Four copies of this exhibit are currently circulating among the participating school districts.

8. Rock layers of northern New Mexico.—Because so many teachers expressed interest in this topic, project staff developed a traveling exhibit on the rock layers of northern New Mexico. The kit includes a large graphic that depicts many of northern New Mexico's rock layers in accurate colors and shapes, along with a collection of rock specimens and fossils found in each layer. Three copies are now circulating among project schools. In addition, each participating school has a fossil kit consisting of locally collected fossil specimens, information about the specimens, and ideas and activities on how to use them in the classroom. This kit has been utilized heavily by the museum's Outreach Program.

9. As a result of participation in NMRSEP, each participating school has instituted a yearly field trip to the New Mexico Museum of Natural History.

10. The Jicarilla Apache Tribe loaned an important fossil skeleton for exhibit in the museum. Students from Cuba Elementary School donated natural history specimens to the museum.
11. NMRSEP staff introduced rural teachers to naturalists and scientists living in their communities who are willing to conduct educational programs in schools. These science subject matter experts included personnel from the New Mexico Bureau of Mines, U.S. Forest Service, Soil Conservation Service, New Mexico Department of Game and Fish, and Game and Fish, Jicarilla Apache Tribe.
12. Students at three of the five schools developed projects focusing on the natural history of their communities, which were displayed at the New Mexico Museum of Natural History.

CONCLUSION

The New Mexico Rural Science Education Project provides an educational model for natural history museum outreach. NMRSEP has demonstrated that, given the commitment to do so, natural history museums can use their unique resources and skills to significantly benefit science education in rural schools. Because both museums and rural centers exist in many other largely rural states, it is a model worth testing elsewhere.

Over the course of the 1988–89 school year, project staff will visit 10 school districts and conduct a total of 16 hours of natural science workshops focusing on the local environment. Project staff will be assisted in these workshops by local teachers who had participated in SESTI. Funding for the workshops and subsequent curriculum development has been furnished by a grant of $58,000 from the New Mexico Commission on Higher Education and the U.S. Department of Education and a fee-for-service of $2000 per school district.

LITERATURE CITED

Gottfried, J. and R. Smith. July, 1988. *The Community as Science Laboratory: Teaching Elementary Science Using Local Natural Resources.* Unpublished manuscript, New Mexico Museum of Natural History, Albuquerque.

APPENDIX. —*Evaluation Instrument Used With Hatch and Magdalena.*
Evaluation of first year of New Mexico Rural Science Education Project.

Please circle one number for each statement.
A (1) =STRONGLY AGREE
a (2) = slightly agree
d (3) = slightly disagree
D (4) = STRONGLY DISAGREE

	A	a	d	D
1. It took most of the year to figure out what was expected of me.	1	2	3	4
2. I enjoyed doing the research on my kit.	1	2	3	4
3. The support from Rural Science Education and Museum people was very good.	1	2	3	4
4. I would have liked a detailed outline of what was needed for putting together a kit.	1	2	3	4
5. I would like established deadlines for completing certain parts of the kit.	1	2	3	4
6. The school system needs to see the value of field trips.	1	2	3	4
7. I really enjoyed the trips the teachers took out in the field.	1	2	3	4
8. This project is helping me get an idea of what to do when I take my students outside.	1	2	3	4
9. If I had access to a computer with a modem, I would use the on-line science network.	1	2	3	4
10. The lack of structure for developing the kits made the work seem overwhelming.	1	2	3	4
11. Developing a kit takes too much time.	1	2	3	4
12. In my kit, I tried to develop activities for several grades.	1	2	3	4
13. It helps to watch someone else present activities in science.	1	2	3	4
14. I would like to have someone come and work with my students on activities directly related to my curriculum.	1	2	3	4
15. I feel that working on the kits was very worthwhile.	1	2	3	4
16. I would be interested in seeing a videotape of different activities that could be used in the classroom.	1	2	3	4
17. This year I would like to work on setting up a science resource room.	1	2	3	4
18. I am more enthusiastic about science since being in this project.	1	2	3	4
19. Sometimes I felt there was a lack of communication between our teachers and the project personnel.	1	2	3	4
20. I have tried to incorporate things I have done in the project into my classroom activities.	1	2	3	4
21. I would like help in developing activities to meet the state curriculum requirements for competencies.	1	2	3	4

22. I would like to see the kits made available to all teachers.   1 2 3 4

23. Working with this project helped me be more aware of the
features of our local environment.   1 2 3 4

24. I find it hard to devote time after school to working on the kit. 1 2 3 4

25. Sometimes the project personnel mentioned doing things
that they never followed up.   1 2 3 4

26. I prefer in-depth information about specific areas of natural
history rather than a little information about broad areas.   1 2 3 4

27. I would like help developing a method of testing hand-on
activities other than a written test.   1 2 3 4

28. I spend more time now having my students do hands-on
activities than I did before being involved in this project.   1 2 3 4

29. I feel that working with the project has made me a better
science teacher.   1 2 3 4

30. I take my students outside to study the natural environment
more than I did before working with this project.   1 2 3 4

PLEASE FEEL FREE TO WRITE ANY COMMENTS OR EXPLANATIONS IN
THE MARGINS OR ON THE BACK OF THE PAGE. THANK YOU!

The Future

Natural History Museums: Directions for Growth
Paisley S. Cato and Clyde Jones, editors
Texas Tech University Press, Lubbock, 1991, iv+252 pp.

FOUNDATION FUNDING
FOR NATURAL HISTORY MUSEUMS

Jerry R. Choate

Abstract.—Nearly all natural history museums need more money than they have. Nevertheless, most museums overlook foundations as a potential source of supplemental funding. Foundations are of two broadly overlapping kinds: corporate foundations and private foundations. Information about corporate foundations, many of which provide grants to museums, can be obtained from several published references. The best source of information about private foundations is The Foundation Center, an independent service organization that provides information on private philanthropic giving. This paper explains how and where to obtain information from The Foundation Center for use in applying for funding from private foundations.

Museum professionals return home from meetings and workshops with their minds full of new ideas. However, at most museums a shortage of ideas is not the most pressing problem. Rather, there is a lack of time and personnel to implement ideas. Time and personnel both translate into money; therefore, what most museums really need is more money. How can museums obtain the additional money they need?

One way, which I have found successful, is to acquire support from a foundation. Persons at large natural history museums already are aware of foundations as an important source of financial support and may not benefit from this information. In fact, some large natural history museums literally would have to close their doors were it not for support from foundations. At the other extreme, many relatively small natural history museums have never received foundation support and could do a much better job of what they do if they could break into this potentially important source of funding. This paper is addressed to persons at the latter museums.

Foundations can be classified in two broadly overlapping categories: corporate foundations and private foundations. Corporate foundations may be large and bureaucratic; the Ford Foundation, for example, recently reported assets totaling 4.7 billion dollars and an annual payroll of 22.6 million dollars. Some corporate foundations deal in large sums of money. For example, the Ford Foundation recently awarded $400,000 to the New Jersey Science and Technology Center. In this regard, several years ago I was a member of a committee that included the directors of numerous natural history museums. During a committee meeting, one of the participants was called out to take an important telephone call. When he returned, smiling, he explained that the call was from the administrator of a corporate foundation informing him that his request for support totaling more than one million dollars had been approved. Grants of this nature, in the tens or hundreds of thousands or even millions of dollars, are provided by a few well-funded corporate foundations to certain large public or private museums as one way of showing the concern of the corporations for the welfare of the public.

There are numerous sources of published information on corporate foundations. In fact, a museum administrator easily can fill a bookcase with expensive volumes on how and to whom to apply for corporate foundation grants. Examples of two such references are *Corporate Foundation Profiles,* which costs $75, and *The Foundation Directory* and *Supplement,* which together cost $120. These and various other references provide information on grant-making policies, financial data, recent grants, and application guidelines for the largest corporate foundations. The volumes are out of date within a year or two of publication, so periodic updates must be purchased. However, they are essential if one wishes to apply for corporate foundation support, but lacks first-hand knowledge about the foundations.

The likelihood that a natural history museum will receive corporate foundation funding is poor unless the museum has exceptional or unique public programs to tout. Larger museums thus have a much better chance than smaller museums of obtaining funding from this source. Small natural history museums might do better with the other category of foundations that I alluded to earlier: private foundations.

Private foundations typically are established by individuals or families, often in wills. The largest of the private foundations (for example, the W. K. Kellogg Foundation Trust) have assets of billions of dollars and award grants each year totaling millions of dollars. These very large private foundations, like corporate foundations, are not apt to fund small natural history museums. At the other extreme are the smallest private foundations, which have assets of just a few thousand dollars and make infrequent small awards. The vast majority of private foundations fall between these extremes and are a potentially important source of support for small museums. There are thousands of these medium-sized private foundations, and few of them, if any, are listed in the references that I mentioned earlier.

So, how can you find out about the existence of private foundations? The best source is The Foundation Center, which is an independent service organization established by foundations to provide information on private philanthropic giving. The Foundation Center disseminates its information on private foundations in various ways, including a national network of library reference collections for free public use. The library reference collections contain sets of Internal Revenue Service form 990-PF for private foundations. Complete sets of these IRS forms for all private foundations are available at the two largest library reference collections, in New York and Washington, DC. IRS forms for all private foundations in the western states can be examined at the library reference collection in San Francisco, and IRS forms for all private foundations in the midwestern states are on file in the Cleveland library reference collection. (See the appendix for addresses and telephone numbers of these library reference collections.)

In addition to these library reference collections, The Foundation Center has cooperating library collections that house information on all private foundations in their respective states or regions. For example, IRS forms for all private foundations in Kansas are available at the Topeka and Wichita public libraries. If you want information regarding private foundations in Kansas or any other state, you can simply visit one of The Foundation Center's cooperating collections in that state and review the IRS forms. Addresses and telephone numbers of cooperating collections containing information on private foundations in their particular states or regions are appended to this paper.

The kinds of information available on private foundations from these sources include: the name, address, and telephone number of the trustee; the names of other members of the board; the assets of the foundation; the total annual gifts of the foundation; the amounts of the largest and smallest gifts; how many gifts the foundation made during the year; the funding priorities of the foundation; any restrictions imposed by the foundation for funding (for example, some

foundations only fund projects in a particular city or county);
information on how to apply for funding; and various additional kinds of information (for example, whether or not
there are deadlines for proposals).

So, what is the likelihood that a small natural history
museum might obtain funding from a private foundation? I
believe that the odds are good and cite some figures from my
experience with private foundations in Kansas.

In terms of foundations, Kansas is a poor state. Many of its
inhabitants depend either directly or indirectly on agriculture for their livelihood, and persons involved in agriculture
seemingly seldom are wealthy. There are a few enclaves of
business entrepreneurship in Kansas, but not many compared
with most other states. Therefore, the number of persons in
Kansas who are sufficiently wealthy to establish philanthropic
foundations is low. Nevertheless, in 1985 there were 331
private foundations in Kansas with combined assets of nearly
200 million dollars. In that year, those foundations made
philanthropic gifts totaling more than 18 million dollars. In
1987, from that list of 331 private foundations, I identified
three dozen or so that potentially might fund the educational
programs of a small natural history museum. Letter
proposals were submitted to the trustees of those foundations, and several of them responded with grants (collectively
totaling more than $27,000) to help with a traveling exhibi-
tion program. This might not seem like much money to persons at large public or private museums, but for small natural
history museums it can make the difference between having a
traveling exhibition program and not having one.

Public education, as exemplified by our traveling exhibition program, is just one of the kinds of museum activities
that can be supported by funding from foundations. One of
my curators regularly receives foundation money in support
of his research. Another curator is negotiating with several
businessmen about establishing a foundation for the purpose
of funding an endowed chair in the museum. Any museum
would have numerous projects for which foundation funding
might be appropriate.

Foundation funding, therefore, is a potential source of
financial support that has been overlooked by most natural
history museums, especially small museums and university
museums. If your museum falls into this category, then my

advice is: first, consider how you might use and justify foundation support; and second, begin discussing your proposed
projects with foundation trustees. Good luck!

APPENDIX.—*The Foundation Center network.*

Listed below are The Foundation Center's national reference collections and
cooperating library collections at which information on private foundations is
available. The lists were modified from those prepared by Rhodes (1989, The
directory of Kansas Foundations, Topeka Public Library, Topeka, KS, 385 pp.).
Because the collections differ with respect to hours open, materials available, and
services provided, you should telephone the collection in which you wish to work.
To check on new locations or current information, call 1-800-424-9836.

A.—*Reference collections operated by The Foundation Center.*

The Foundation Center
8th Floor
79 Fifth Avenue
New York, NY 10003
(212) 620-4230

The Foundation Center
1001 Connecticut Avenue, NW
Washington, DC 20036
(202) 331-1400

The Foundation Center
Kent H. Smith Library
1442 Hanna Building
Cleveland, OH 44115
(216) 861-1933

The Foundation Center
Room 312
312 Sutter Street
San Francisco, CA 94108
(415) 397-0902

B.—*Cooperating collections that have sets of private foundation information
returns (IRS Form 990-PF) for their states or regions.*

ALABAMA
Birmingham Public Library
2020 Park Place
Birmingham, AL 35203
(205) 226-3600

Auburn University at Montgomery
  Library
I-85 at Taylor Road
Montgomery, AL 36193
(205) 271-9649

ALASKA
University of Alaska, Anchorage Library
3211 Providence Drive
Anchorage, AK 99508
(907) 786-1848

ARIZONA
Phoenix Public Library
Business and Sciences Department
12 East McDowell Road
Phoenix, AZ 85257
(602) 262-4636

Tucson Public Library
200 South Sixth Avenue
Tucson, AZ 85726
(602) 791-4393

ARKANSAS
Westark Community College Library
5210 Grand Avenue
Fort Smith, AR 72913
(501) 785-7000

Central Arkansas Library System
Reference Services
700 Louisiana Street
Little Rock, AR 72201
(501) 370-5950

CALIFORNIA
California Community Foundation
Funding Information Center
3580 Wilshire Blvd., Suite 1660
Los Angeles, CA 90010
(213) 413-4042

Community Foundation for Monterey
  County
420 Pacific Street
Monterey, CA 93940
(408) 375-9712

San Diego Community Foundation
525 "B" Street, Suite 410
San Diego, CA 92101
(619) 239-8815

Orange County Community
  Developmental Council
1695 W. MacArthur Blvd.
Costa Mesa, CA 92626
(714) 540-9293

Peninsula Community Foundation
1204 Burlingame Avenue
Burlingame, CA 94011
(415) 342-2505

Santa Barbara Public Library
40 East Anapamu
Santa Barbara, CA 93102
(805) 962-7653

COLORADO
Denver Public Library
Sociology Division
1357 Broadway
Denver, CO 80203
(303) 571-2190

CONNECTICUT
Hartford Public Library
Reference Department
500 Main Street
Hartford, CT 06103
(203) 293-6000

DELAWARE
Hugh Morris Library
University of Deleware
Newark, DE 19717
(302) 451-2965

FLORIDA
Jacksonville Public Library
Business, Science & Documents
122 North Ocean Street
Jacksonville, FL 32206
(904) 630-2665

Miami-Dade Public Library
Humanities Department
101 W. Flagler Street
Miami, FL 33130
(305) 375-2665

Orange County Library System
101 E. Central Blvd.
Orlando, FL 32801
(407) 425-4694

Leon County Public Library
Funding Resource Center
1940 North Monroe Street
Tallahassee, FL 32303
(904) 478-2665

GEORGIA
Atlanta-Fulton Public Library
Ivan Allen Department
1 Margaret Mitchell Square
Atlanta, GA 30303
(404) 730-1700

HAWAII
Thomas Hale Hamilton Library
University of Hawaii
2550 The Mall
Honolulu, HI 96822
(808) 948-7214

IDAHO
Boise Public Library
715 S. Capitol Blvd.
Boise, ID 83702
(208) 384-4466

Caldwell Public Library
1010 Dearborn Street
Caldwell, ID 83605
(208) 459-3242

ILLINOIS
Donors Forum of Chicago
53 W. Jackson Blvd., Room 430
Chicago, IL 60604
(312)431-0265

Evanston Public Library
1703 Orrington Avenue
Evanston, IL 60201
(312) 866-0305

Sangamon State University Library
Shepherd Road
Springfield, IL 62794
(217) 786-6633

INDIANA
Allen County Public Library
900 Webster Street
Fort Wayne, IN 46802
(219) 424-7241

Indianapolis-Marion County Public
  Library
40 East St. Clair Street
Indianapolis, IN 46206
(317) 269-1733

IOWA
Public Library of Des Moines
100 Locust Street
Des Moines, IA 50308
(515) 283-4259

KANSAS
Topeka Public Library
1515 West Tenth Street
Topeka, KS 66604
(913) 233-2040

Wichita Public Library
223 South Main
Wichita, KS 67202
(316) 262-0611

KENTUCKY
Louisville Free Public Library
Fourth and York Streets
Louisville, KY 40203
(503) 561-8600

LOUISIANA
East Baton Rouge Parish Library
Centroplex Branch
120 St. Louis Street
Baton Rouge, LA 70802
(504) 389-4960

New Orleans Public Library
Business and Science Division
219 Loyola Avenue
New Orleans, LA 70140
(504) 596-2580

Shreve Memorial Library
424 Texas Street
Shreveport, LA 71120
(318) 226-5894

MAINE
University of Southern Maine
Office of Sponsored Research
246 Deering Avenue, Room 628
Portland, ME 04103
(207) 780-4871

MARYLAND
Enoch Pratt Free Library
Social Science and History
  Department
400 Cathedral Street
Baltimore, MD 21201
(301) 396-5320

MASSACHUSETTS
Associated Grantmakers of
  Massachusetts
294 Washington Street, Suite 840
Boston, MA 02108
(617) 426-2608

Boston Public Library
666 Boylston Street
Boston, MA 02117
(617) 536-5400

Western Massachusetts Funding
   Resouce Center
Campaign for Human Development
73 Chestnut Street
Springfield,, MA 01108
(413) 732-3175

Grants Resource Center
Worcester Public Library
Salem Square
Worcester, MA 01608
(508) 799-1655

MICHIGAN
Alpena County Library
211 North First Avenue
Alpena, MI 49707
(517) 356-6188

Henry Ford Centennial Library
16301 Michigan Avenue
Dearborn, MI 48126
(313) 943-2337

Purdy Library
Wayne State University
Detroit, MI 48202
(313) 577-4040

Michigan State University Libraries
Reference Library
East Lansing, MI 48824
(517) 353-8818

Farmington Community Library
32737 West 12 Mile Road
Farmington Hills, MI 48018
(313) 553-0300

University of Michigan-Flint Library
Reference Department
Flint, MI 48502
(313) 762-3408

Grand Rapids Public Library
Business Department
60 Library Plaza NE
Grand Rapids, MI 49503
(616) 456-6300

Michigan Technological University
   Library
U.S. Highway 41
Houghton, MI 49931
(906) 487-2507

Sault Ste. Marie Area Public Schools
Office of Compensatory Education
460 W Spruce Street
Sault Ste. Marie, MI 49783
(906) 635-6619

MINNESOTA
Duluth Public Library
520 Superior Street
Duluth, MN 55802
(218) 723-3802

Southwest State University Library
Marshall, MN 56258
(507) 537-7278

Minneapolis Public Library
Sociology Department
300 Nicollet Mall
Minneapolis, MN 55401
(612) 372-6555

MISSISSIPPI
No cooperating collections that have
sets of IRS Form 990-PF for private
foundations are listed for Mississippi.

MISSOURI
Clearinghouse for Midcontinent
   Foundations
University of Missouri, Kansas City
Law School, Suite 1-300
52nd Street and Oak
Kansas City, MO 64113
(816) 276-1176

Kansas City Public Library
311 East 12th Street
Kansas City, MO 64106
(816) 221-9650

Metropolitan Association for
   Philanthropy, Inc.
5585 Pershing Avenue, Suite 150
St. Louis, MO 63112
(314) 361-3900

Springfield-Greene County Library
397 East Central Street
Springfield, MO 65801
(417) 866-4636

MONTANA
Eastern Montana College Library
1500 N 30th Street
Billings, MT 59101
(406) 657-1662

Montana State Library
Reference Department
1515 E 6th Avenue
Helena, MT 59620
(406) 444-3004

NEBRASKA
University of Nebraska
106 Love Library
14th and R Streets
Lincoln, NE 68588
(402) 472-2848

W. Dale Clark Library
Social Sciences Department
215 South 15th Street
Omaha, NE 68102
(402) 444-4826

NEVADA
Las Vegas-Clark County Library District
1401 East Flamingo Road
Las Vegas, NV 89109
(702) 733-7810

Washoe County Library
301 South Center Street
Reno, NV 89501
(702) 785-4012

NEW HAMPSHIRE
New Hampshire Charitable Fund
One South Street
Concord, NH 03301
(603) 225-6641

NEW JERSEY
New Jersey State Library
Governmental Reference
185 West State Street
Trenton, NJ 08625
(609) 292-6220

NEW MEXICO
New Mexico State Library
325 Don Gaspar Street
Sante Fe, NM 87505
(505) 827-3824

NEW YORK
New York State Library
Cultural Education Center
Humanities Section
Empire State Plaza
Albany, NY 12230
(518) 474-5161

Buffalo and Erie County Public Library
Lafayette Square
Buffalo, NY 14202
(716) 858-7103

Levittown Public Library
One Bluegrass Lane
Levittown, NY 11756
(516) 731-5720

Plattsburgh Public Library
15 Oak Street
Plattsburgh, NY 12901
(518) 563-0921

Rochester Public Library
Business Division
115 South Avenue
Rochester, NY 14604
(716) 428-7328

Onondaga County Public Library at
the Galleries
447 S Salina Street
Syracuse, NY 13202
(315) 448-4636

White Plains Public Library
100 Martine Avenue
White Plains, NY 10601
(914) 682-4480

Suffolk Cooperative Library System
627 N Sunrise Service Road
Bellport, NY 11713
(516) 286-1600

NORTH CAROLINA
Asheville-Buncomb Technical
    Community College
Learning Resource Center
340 Victoria Road
Asheville, NC 28801
(704) 254-1921, ext. 300

The Duke Endowment
200 S Tryon Street, Ste. 1100
Charlotte, NC 28202
(704) 376-0291

North Carolina State Library
109 East Jones Street
Raleigh, NC 27611
(919) 733-3270

The Winston-Salem Foundation
229 First Union Bank Building
Winston-Salem, NC 27101
(919) 725-2382

NORTH DAKOTA
The Library North Dakota State
    University
Fargo, ND 58105
(701) 237-8886

OHIO
Public Library of Cincinnati and
    Hamilton County
Education Department
800 Vine Street
Cincinnati, OH 45202
(513) 369-6940

Dayton and Montgomery County
    Public Library
Grants Information Center
215 E Third Street
Dayton, OH 45402
(513) 227-9500, ext. 211

Toledo-Lucas County Public Library
Social Science Department
325 Michigan Street
Toledo, OH 43623
(419) 259-5245

OKLAHOMA
Oklahoma City University Library
2501 N Blackwelder
Oklahoma City, OK 73106
(405) 521-5072

Tulsa City-County Library System
400 Civic Center
Tulsa, OK 74103
(918) 596-7944

OREGON
Multnomah County Library
Government Documents Room
801 SW Tenth Avenue
Portland, OR 97205
(503) 223-7201

Pacific Non-Profit Network
Grantsmanship Resource Library
33 N Central, Ste. 211
Medford, OR 97501
(503) 779-6044

PENNSYLVANIA
Erie County Public Library
3 South Perry Square
Erie, PA 16501
(814) 451-6927

Dauphin County Library System
101 Walnut Street
Harrisburg, PA 17101
(717) 234-4961

The Free Library of Philadelphia
Logan Square
Philadelphia, PA 19103
(215) 686-5423

Hillman Library
University of Pittsburgh
Pittsburgh, PA 15260
(412) 648-7722

RHODE ISLAND
Providence Public Library
Reference Department
150 Empire Street
Providence, RI 02903
(401) 521-7722

SOUTH CAROLINA
Charleston County Library
404 King Street
Charleston, SC 29403
(803) 723-1645

South Carolina State Library
Reference Department
1500 Senate Street
Columbia, SC 29211
(803) 734-8666

SOUTH DAKOTA
South Dakota State Library
800 Governors Drive
Pierre, SD 57501
(605) 773-3131
(800) 594-1841 (South Dakota
residents)

TENNESSEE
Knoxville-Knox County Public Library
500 West Church Avenue
Knoxville, TN 37902
(615) 544-5750

Memphis & Shelby County Public
   Library
1850 Peabody Avenue
Memphis, TN 38104
(901) 725-8876

Public Library of Nashville and
   Davidson County
8th Avenue, N and Union Street
Nashville, TN 37211
(615) 259-6256

TEXAS
Hogg Foundation for Mental Health
University of Texas
Austin, TX 78713
(512) 471-5041

Corpus Christi State University Library
6300 Ocean Drive
Corpus Christi, TX 78412
(512) 994-2608

El Paso Community Foundation
201 E Main
El Paso, TX 79901
(915) 533-4020

Texas Christian University Library
Funding Information Center
Ft. Worth, TX 76129
(817) 921-7664

Houston Public Library
Bibliographic Information Center
500 McKinney Avenue
Houston, TX 77002
(713) 236-1313

Lubbock Area Foundation
502 Commerce Bank Building
Lubbock, TX 79401
(806) 762-8061

Funding Information Library
507 Brooklyn
San Antonio, TX 78215
(512) 227-4333

Dallas Public Library
Grants Information Service
1515 Young Street
Dallas, TX 75201
(214) 670-1487

Pan American University Learning
   Resource Center
1201 W University Drive
Edinburg, TX 78539
(512) 381-3304

UTAH
Salt Lake City Public Library
Business and Science Department
209 East Fifth South
Salt Lake City, UT 84111
(801) 363-5733

VERMONT
Vermont Department of Libraries
Reference Services
109 State Street
Montpelier, VT 05602
(802) 828-3268

VIRGINIA
Hampton Public Library
Grants Resources Collection
4207 Victoria Blvd
Hampton, VA 23669
(804) 727-1154

Richmond Public Library
Business, Science, and Technology
101 East Franklin Street
Richmond, VA 23219
(804) 780-8223

WASHINGTON
Seattle Public Library
1000 Fourth Avenue
Seattle, WA 98104
(206) 386-4620

Spokane Public Library
Funding Information Center
West 906 Main Avenue
Spokane, WA 99201
(509) 838-3364

WEST VIRGINIA
Kanawha County Public Library
123 Capital Street
Charleston, WV 25304
(304) 343-4646

WISCONSIN
Marquette University Memorial Library
1415 West Wisconsin Avenue
Milwaukee, WI 53233
(414) 288-1515

University of Wisconsin-Madison
    Memorial Library
728 State Street
Madison, WI 53706
(608) 262-3242

WYOMING
Laramie County Community College
    Library
1400 East College Drive
Cheyenne, WY 82007
(307) 778-1205

Natural History Museums: Directions for Growth
Paisley S. Cato and Clyde Jones, editors
Texas Tech University Press, Lubbock, 1991, iv+252 pp.

THE SECOND CENTURY: ANTHROPOLOGY IN NATURAL HISTORY MUSEUMS

P. Lynn Denton

Abstract.—The founding and establishment of modern anthropological research in the 19th-century United States fundamentally was guided through affiliation with major natural history museums, notably the Smithsonian Institution. In these settings, the early systematic collections and milestone descriptive publications were produced as part of the broader effort to gather data on all aspects of the continent's natural and physical characteristics. Thus, the prehistoric and ethnographic record of human inhabitants was linked organizationally with all other forms of natural life under study. After the turn of the century, the center of anthropological research and activity shifted from natural history museums to university departments where it has remained. As shifts in theoretical and research orientation of the field occurred, anthropological natural history collections were often ignored. Simultaneously, issues of worldwide and Native American patrimony challenged the appropriateness of the anthropological collections in our institutions.

These challenges have gained insistence in the recent establishment of a Native American Indian museum that is separate from the Smithsonian's National Museum of Natural History. How do anthropological research concerns of the late 20th century and early 21st century apply to fundamental collecting, research, and interpretation functions of museums? Do natural history museums or university natural history museums specifically have a unique but undefined role in anthropology for the future?

Over 20 years ago, a series of papers and reports by museum anthropologists Fenton (1960), Dockstader (1967), Collier and Tschopik (1954), and others lamented the reduced volume and impact of museum-based anthropological research. Subsequent articles (Sturtevant, 1969) held such studies were never a major segment of the field, but acknowledged that the range and number of such efforts had decreased proportionally. Variations of these concerns remain as museum anthropologists continue to grapple with perceptual patterns rooted in anthropology of the Victorian era (Rathje, 1982; Cantwell and Rothschild, 1981). Several historical factors are significant in evaluating the current situation as well as future directions.

The founding and establishment of American anthropology in the 19th century was integrally linked to early U.S. natural history museums. This initial museum period 1840–1890 (Sturtevant, 1969) or 1869–1900 (Fenton, 1960) in anthropology included the Smithsonian Institution, which was connected with the Bureau of American Ethnology, as well as the American Museum of Natural History, Field Columbian Museum of Chicago, and university-based Peabody Yale, Peabody Harvard, and the University of Pennsylvania museums. The specific significance of each institution's role in developing early anthropological theory and methodology varies per author; nevertheless, in the most general terms, anthropological research during this time was strongly descriptive and comparative in nature. In all four subfields—linguistics, physical anthropology, archeology, and ethnology—early efforts were focused on recording and collecting materials and information "preparatory to . . . deductions and generalization" (Hinsley, 1981). These activities paralleled the data gathering and collecting for the continent in the physical and biological sciences. Joseph Henry, as first director of the Smithsonian Institution, exerted great influence on the young field through his conviction that the application of the research standards, methodological systems, and accurate instrumentation of the physical sciences was essential to remove anthropology from a "hobbyist" status. His field researchers were urged to develop a rigorous, inductive anthropology program based on a system of observation and thereby free from biases and speculation (Hinsley, 1981). Sturtevant (1969) has maintained that despite this typological and classificatory emphasis, "museum collections were only marginally related to the development of theories of cultural evolution which was the main focus of interest of anthropology during this period." However, the "basic pattern of anthropological activities in American museums was well established. These activities consisted of programs of exhibition, research, scientific and popular publications, contributions to journals, teaching and popular lectures (Collier and Tschopik, 1954)." The great systematic collections were established, seminal descriptive and historical publications produced, and exhibitions developed to enlighten and encourage "high minded inquiry" by the viewer.

Anthropology's beginning as part of natural history museums affected not only its concepts and its methods, but also its personnel and their outlook (Fenton, 1960). Because there were no formal programs of training, anthropologist generally were derived from other natural history disciplines. Thus, it is not surprising that they not only borrowed from the methods of natural history disciplines, but that they also emphasized the place of anthropology in "natural science or as a branch of natural history" (Collier and Tschopik, 1954; Osgood, 1979). As a case in point, at Texas Memorial Museum (a university natural history institution), the first Paleoindian research and collections in the state were generated by a museum geologist.

Under the leadership of Franz Boas at Columbia, major research initiatives and energy had shifted to university departments by World War I (Hinsley, 1981), and as Fenton (1974) stated, whereas anthropology was nurtured in museums it matured in universities. During this period, primary employment in anthropology shifted from museums to universities as did the financial support for fieldwork. Theoretical research in ethnology abandoned a descriptive and collection foundation to focus on social and mental culture. By the middle of this century, the reduced concern for material culture oriented research resulted in greater numbers of anthropology curators with responsibilities for collections where they had no research ties or interest. Physical anthropology research also grew in noncollection-based fields such as genetics and primate ethology. The lack of utilization, care, and protection for collections that was accentuated by these trends prompted such reports as the Council for Museum Anthropology/National Science Foundation's *Systematic Research Collections in Anthropology* (Ford 1977), which described the irreplaceable nature of collection resources and evaluated their current status in 1977. Archeology alone retained its essential collection-oriented focus augmented by theoretical and methodological advances. At the same time, increased Native North American political participation and awareness challenged the legal and ethical appropriateness of museum-held collections. Additionally, campaigns to separate Native North American anthropological materials from natural history institutions have gained strength. Interestingly enough, Fenton (1960) and Sturtevant (1969) presented and discussed this same separatist view in the 1960s.

At that same time, several questions and proposed solutions were formulated for reintroducing museum-based anthropological research into the forefront of theoretical inquiry. First, the potential contributions of material culture studies were reaffirmed suggesting such areas for special attention as culture change. In a 1972 article, John and Jocelyn Mori (1972) suggested "that museum objects serve as a body of data against which a myriad of anthropological assumptions [can] be tested as well as a source from which to draw generalizations concerning many facets of human behavior." The validity of this position has been demonstrated in the New York Academy of Sciences 1981 conference and publication in which over 30 papers addressed (1) new or reformulated questions of older collections, (2) new scientific techniques for existing data or, (3) solutions to problems in modern research on older collections (Cantwell 1981). Material culture studies have received greater interest and support in the last decade as expressed in published and presented papers and student theses. Improvements in formulation of research goals and methods will have to continue for as Sturtevant (1969) stated, "what causes shifts in research emphasis is the discovery of quite specific problems and methods . . . that promise advances clearly related to other important interests of the discipline." Thus the study of the physical evidence of human behavior seems to hold continued promise and validity within a natural history museum setting. In 1965, the American Anthropological Association established the Committee on Anthropology Research in Museums, which provided approximately 40 Wenner Gren funded museum research fellowships over several years. The program was conceived as the "best way to improve a bad situation and to accomplish our aims . . . to train a new generation of scholars who would use collections as resources for research both in the museum and in the field" (Fenton, 1974). In a recent personal communication with Fenton regarding the long term results of this effort, he noted substantial achievement in the excellent quality of current material culture research and in the number of fellowship participants who are now leading professionals in the field. However, it must be recognized that these individuals are still few in number and that the sustained development of material culture research interest will rely on strengthening departmental and museum relationships and opportunities.

Second, exhibitions and related symposia and publications were proposed as vehicles for contributing to contemporary anthropological theory. The National Endowment for the Humanities and other granting agencies have played a primary role in recent years by providing assistance for scholarly-based endeavors that answer needs in the field as well as meeting public obligations. Although many outdated culture area exhibits have yet to be revitalized, increased numbers of museums are seeking current conceptual interpretations versus more traditional technological or stylistic formats; the "omnium gatherum" that Joseph Henry so feared. Exhibit related research, however, has yet to receive the peer recognition of other research products.

Third, to charges that museums still approach culture in a fragmentary and often "freeze frame" format, it must be countered that the specializations within ethnology must seek improved integration between mental, material, and social culture. For example, the current research in expressive culture and semiotics poses many unrealized opportunities for museum application.

Fourth, the assertion that the functions of collection research, maintenance, and interpretation could not be addressed by a single individual (Fenton, 1960; Sturtevant, 1969; Crompton, 1968) has been partially remedied through training of collections management staff and collection care specialists. A concomitant increase in research output has not been forthcoming, however.

Despite the above, other aspects of the role of anthropology in natural history museums for the next two decades and into the new century are yet to be determined. Natural history museums are often actively involved in issues of biological diversity and destruction. What are our responsibilities regarding the issues of contemporary cultural diversity and destruction? The Victorian search for order and control (Hinsley, 1981) generated museums as institutions that embodied an imposed order, ". . . as a bastion of certainties . . . as an important defense . . . against a world of more human variety than previously imagined . . ." (Hinsley, 1981). These perceptions are still with anthropologists in natural history museums. Kreps (1988) has recently discussed a Dutch example of "decolonizing" anthropology museums from this Victorian heritage in order that a broader range of contemporary cultural issues, international understanding and perspective, and cross-cultural relations may be addressed in programming.

There is general acknowledgment in most of the subfields that a dwindling resource base—whether fewer archeological

sites or politically restricted ethnographic field work—will increase the value of existing collections and data for the future
(Cantwell and Rothschild, 1981). Archeologists and physical
anthropologists particularly have recognized this limitation.
The collections will remain as "cultural vouchers" and as essential in many ways as the literature (Fenton, 1960; Ford, 1977).

Perhaps the most intriguing future for museum anthropology may lie in the impact of the "information age." Information-age technology is changing not only what is collected but
the definitions of value and significance and the process for
decision making.

MacDonald (1988) proposed that information-age values
are shifting from objects as "rare wealth . . . in a commodity
based system of the past" to essential icons in an experience
based system. Indicative of this trend is Disney Corporation's
establishment of a new Museums Branch that is programming
African art and ethnographic collections as a part of the
corporation's initiative to become cultural brokers in addition to entertainment brokers (MacDonald, 1987). This
same technology also enables previously inconceivable types
and amounts of data to be collected and retained. Collier
and Tschopik (1954) foresaw an important role for museum
anthropology in this regard through active collection of specialized language, music, photographic, and motion picture
records in the mid-1950s.

In conclusion, the efforts to re-establish greater professional and student acknowledgment and interest for collec-
tion-based research have achieved measurable success.
Advances have been made in presenting significant exhibits
based on contemporary anthropological research issues, however, support for these endeavors must be continued and
broadened at all levels of interpretive planning and implementation. The ability to approach cultures holistically in
a museum environment will depend on the sharing of an integrated perspective, especially between those within the sub-
disciplines' highly specialized areas and museum staff. The
debate over the position of anthropology within natural history will intensify and may result in shifts to independent institutional standing. Finally, the training and employment of
collection-care specialists appears to be alleviating some of
the multiple job pressures that were attributed to the reduc-
tion in research efforts in the past. This improved collection

care, combined with a renewed emphasis on the essential role of museum research, portend well for the future.

With regard to the next century, a critical role for museum anthropology can be postulated in synthesizing information age technologies with traditional research, collection, and interpretive functions. The integration of these capabilities with the diversity of anthropological knowledge in the unique setting of the museum can certainly forge exciting new contexts for cultural understanding.

LITERATURE CITED

Cantwell, A. E. and N. A. Rothschild. 1981. The future of the past. Pp.579-583, *in* The research potential of anthropological museum collections (Cantwell, A.M. and J.B. Griffin, eds.) New York Academy of Sciences Annuals, 371:1-583.

Collier, D. and H. Tschopik, Jr. 1954. The role of museums in American anthropology. Amer. Anthropologist, 56(5, Part 1):768-779.

Crompton, A. W. 1968. The present and future course of our museum. Museum News, 46(5):35-37.

Dockstader, F. J. 1967. Anthropology and the museum. Pp. 132-142, *in* The Philadelphia Anthropological Society, papers presented on its golden anniversary (J. W. Gruber, ed.). Temple Univ. Publications, New York, 162 pp.

Fenton, W. N. 1960. The museum and anthropological research. Curator, 3:327-355.

———. 1974. The advancement of material culture studies in modern anthropological research. Pp.15-36, *in* The human mirror (M. Richardson, ed.) Louisiana State Univ. Press, Baton Rouge, 365 pp.

Ford, R. I. 1977. Systematic research collections in anthropology an irreplaceable national resource. Peabody Museum, Harvard Univ., 81 pp.

Hinsley, C. M., Jr. 1981. Savages and scientists. The Smithsonian Institution and the development of American anthropology 1846-1910. Smithsonian Institution Press, Washington, D.C., 319 pp.

Kreps, C. 1988. Decolonizing anthropology museums: The Tropenmuseum, Amsterdam. Museum Studies Journal, 3(2):56-63.

MacDonald, G. F. 1987. The future of museums in the Global Village. Museum, 155:209-216.

———. 1988. The future of museums in the Global Village. Museum News, 166(7):69-71.

Mori, J. L. and J. I. Mori. 1972. Revising our conceptions of museum research. Curator, 15:189-199.

Osgood, C. 1979. Anthropology in museums of Canada and the United States. Milwaukee Public Museum, Milwaukee, 109 pp.

Rathje, W. L. 1982. Modern material culture studies. Pp.647-683 *in* Advances in archaeological method and theory. (M. B. Schiffer, ed.) Academic Press, New York, 690 pp.

Sturtevant, W. C. 1969. Does anthropology need museums? Proc. Biol. Soc. Washington, 82:619-649.

Natural History Museums: Directions for Growth
Paisley S. Cato and Clyde Jones, editors
Texas Tech University Press, Lubbock, 1991, iv+252 pp.

NATURAL HISTORY IN THE 20TH CENTURY: AN OXYMORON?

Charlotte M. Porter

Like the word curator, natural history is a museum word, but what does it really mean? In 1940, Webster's *New International Dictionary* defined *natural history* as: "Formerly, the study, description, and classification of animals, plants, minerals, and other natural objects, thus including the modern science of zoology, botany, mineralogy, etc., insofar as they existed at that time; now commonly restricted to a study of these subjects in a more or less unsystematic way." Unsystematic systematics? No museum curator today would agree with this definition of natural history, but scholarly tradition reinforces it. In the early 17th century, Francis Bacon wrote: "In natural history we see there hath not been that choice and judgment used as ought to have been, as may appear in the writings of Plinius, Cardanus, Albertus, and divers of the Arabians." (Best and Brightman, 1974). With the list, Bacon was citing the best, not the worst, of the scientific sources. Moslem medical authorities preserved ancient Greek and Roman science during the Dark Ages of Europe; Albertus Magnus attempted to reconcile the zoological writings of Aristotle with the teachings of the Catholic Church; Geronimo Cardano was a gifted Renaissance mathematician, fascinated by games of chance; and it was Pliny the Elder whose first century Latin encyclopedia, *Historia Naturalis*, gave to natural history both its name and broad scope.

Pliny's work was, by his own confession, derivative. His volumes initiated the close relationship of bibliography and natural history that 1700 years later would become fundamental to biological systematics and nomenclature (see Meisel, 1924–1929). A busy Roman official, he somehow found time to search through 2000 books for 20,000 facts about nature. The 37 volumes of his compilation that survived served the Christian and Islamic Middle Ages as a great dictionary about nature. Pliny died in the eruption of Mount Vesuvius that destroyed Pompeii in 79 AD. Much copied and frequently cited, Pliny's *Natural History* was a source for the facts and fables that filled popular and often moralistic bestiaries and plant books. In this process, manuscript copies and compendiums became corrupted. Illustrations no longer matched accompanying texts, and the texts themselves became confused or embellished in the retelling: dragons joined the land animals; nereids joined men of the sea. For all its faults, Pliny's work never lost its appeal, however, and it was one of the first books to be printed during the Italian Renaissance. In Book II, Christopher Columbus—or for that matter, any of his literate distractors—could read Pliny's assertion that the world has for its shape "the rounded appearance of a perfect sphere" (Whalley, 1982).

During the first 100 years that followed Columbus' landfall, colorful bestiaries and herbals gave way to finely illustrated books of natural history. Compilations, called "pandects," brought together extracts from older scientific authors with the new knowledge that geographical exploration generated (McLean, 1972). A renewed interest in natural objects was part of an age that witnessed the discovery of the New World and the rebirth of scientific literacy. European naturalists of the 16th-century Renaissance attempted to observe nature directly and to record species on the basis of collected specimens (Debus, 1978). On the other side of the world, the spectacular gardens of Montezuma in Mexico provided a unique impetus for the rapid development of living collections of species totally unknown to ancient Greek and Latin authors. These, of course, had no precedent in Pliny's work. Large-scale menageries and gardens of exotic species were soon the pride of princes and artists of southern Europe. Among the wonders of the Americas, the remarkable birds—turkeys, hummingbirds, toucans, and parrots—impressed early travelers, who returned to Europe with both specimens and fabulous accounts. As a result, by 1516, the great Italian master Raphael had models for the American birds he painted in the Loggetta of the Cardinal of Bibbiena at the Vatican (Hounour, 1975).

Before the close of the 16th century, more public viewing places were open to visitors—cabinets of geographical curiosities. These typically included objects of natural history, American Indian novelties, Chinese porcelain, Roman coins, and other relics. By the mid-17th century, these "cabinets of the curious" had become prototypes for private museums of natural history. A room of Ole Worm's museum in Copenhagen, for example, displayed stuffed quadrupeds and fish, miscellaneous *animalium partes*, roots, and mineral salts in the midst of bows and arrows, paddles, and sundry footwear from around the world. The study of natural history was not yet guided by the 18th-century scientific principles of Carolus Linnaeus, and the idea of the *Wunderkammer* inspired a magnificent series of tapestries called "Les Indes," woven for Louis XIV at the Gobelins factory in 1687. Five years after the French acquisition of Louisiana, these large woven tableaux brought together assemblages of plants and animals—real and imagined—with compelling vividness. The role of the artist in the European rediscovery of natural history was soon supplanted by that of the world traveler.

The grand voyages of circumnavigation launched by France, Britain, and Holland during the late 18th century created the foreign study collections that eventually gave rise to the science of ornithology. The impact of organized exploration can be seen at the National Museum in Paris. The museum, arguably then the finest in the world, contained 463 bird specimens in 1793; that number increased by 3000 within the decade (Farber, 1977). In the young American Republic, Francophiles such as Thomas Jefferson were impressed by this French progress in science (Martin, 1952). The uncovering of large fossil animals also prompted the study of natural history (Simpson, 1942). Indeed, for more than a century, the discovery and accumulation of exotic birds and giant bones, later combined with the Victorian taste for big game hunting, characterized many museum activities in the United States. Early museums in American cities of science, however, followed a different course from their European counterparts (Bell *et al.*, 1967). Public museums preceded the large private collections typical of Europe. The first was Charleston Library Society in 1773 (Bragg, 1923). Charles Willson Peale's more ambitious Philadelphia Museum followed in 1786 (Sellers, 1986; 1953).

Peale's blend of artistry, craft, and science had long-term consequences for public interest in natural history (Ellis, 1966). Apprenticed as a saddlemaker in his youth, Peale studied painting briefly in London (Sellers, 1969; 1947). After the American Revolution, he became a moderately successful portrait painter, and in 1784, began to plan a natural history collection in conjunction with his portrait gallery of

Revolutionary War heroes. An early museum ticket, which sold for 25 cents, shows nature as an open book and reads, "The Birds and Beasts will teach thee!" Located in the Philosophical Hall of the American Philosophical Society from 1794 to 1802, Peale's museum eventually exhibited more than 243 birds and 212 mounted qudrupeds, as well as the famous "mammoth," fishes, shells, rocks, and insects (Peale, *et al.*, 1796). Returning to his early training in saddlemaking, he stretched animal skins over wooden forms to restore lifelike attitudes. Often, he hand-carved the internal limbs for these mounts and molded glass eyes. The museum later moved to the Pennsylvania State House, now Independence Hall. Departing from European custom, Peale displayed bird specimens in glass fronted cases backed by naturalistic views painted by his artistic children. One limitation to museum expansion was the preservation of collections. Neither the brandy nor the embalming spices used to transport and store bird skins warded off destructive insects. Peale surmounted these problems through trial and error and worked out basic methods for taxidermy using arsenic. Despite the hazards of handling arsenic, his method proved effective. About 50 of Peale's mounted birds can be found at the Museum of Comparative Zoology, in Cambridge, Massachusetts, where after almost 200 years, the colorful Chinese pheasants Peale obtained from George Washington are still on display (Faxon, 1915). Other materials once housed at Peale's Museum are now exhibited next door at the Peabody Museum.

Many of Peale's visitors preferred his animal displays to his paintings, but Peale's artistic creativity was not easily discouraged. With Jefferson's endorsement, Peale undertook the management of the first public collections with national, rather than regional, emphasis. Peale also loved new technology. His was the first commercial gallery to install overhead lighting, and he used his museum space for magic lantern shows and musical presentations (Peale, 1800; Flexner, 1954). These social events were well attended, but Peale's undeniable triumph and the one that brought him a modicum of gain was the exhumation of the mammoth from a swamp in Orange County, New York. Peale mounted the entire skeleton in his museum in 1801 and, with the aid of his sons, toured a second skeleton in Europe. Peale's mastodon was the first complete reconstruction of a fossil vertebrate. The giant creature made "mammoth" a household word and established the *sine qua non* for other natural history museums. By 1816, Peale's annual gross receipts of $11,924 indicated a paid attendance of nearly 48,000 people, and Peale's sons attempted to start similar enterprises in Baltimore and New York. Regarding his own motivations, he concluded that "the gratification which every new object produced in the mind of an enthusiastic man is all powerful" (Greene, 1966).

Personal dedication aside, however, Peale realized that the key to posterity was the establishment by Congress of a national museum. His museum failed to realize that goal. Instead the museum became the family institution of employment for the proprietor's many sons and daughters, former slave, numerous in-laws, and the city's overflow of talented banknote engravers. Peale's museum provided other naturalists and artists with specimens to study and draw, and also with an audience eager to buy their works. His imaginative displays created a taste for natural history and instructed more than a generation of visitors before its acquisition by P.T. Barnum in 1849-50 (Betts, 1959; Perry, 1976). In particular, Peale's bird collections permitted the production of Alexander Wilson's splendid bird book (Willson, 1808–13). Wilson's landmark publication (1808–1814), in turn, inspired other Philadelphian naturalists to do original research and to use American presses to publish their works (Porter, 1986). Publication began to play an increasingly important role in American natural history.

From humble beginnings in 1812, the Philadelphia Academy of Natural Sciences organized natural history in the direction of zoology (Phillips, 1953). In addition to the influence of Peale's collections, there were other reasons for this emphasis. By the turn of the century, the Parisian zoologist, Georges Cuvier, had identified fossil species on both sides of the Atlantic. The remains of these vanished animals posed profound questions about nature and creation. As one tourist remarked upon seeing Peale's mounted mastodon, "Perhaps we ought to imagine if Noah found it too large and troublesome to put in the ark, and therefor left the poor animal to perish" (Blane, 1828). On a more serious level, these big bones also attracted political attention. The early western expeditions Jefferson organized were instructed to look for these great animals and their remains (Jackson,

1981). More expeditions accompanied the settling of the American West, but lacking institutional structure no botanical collection emerged as a lasting national herbarium. In contrast, great zoological and ethnographic collections were amassed by the National Museum of the Smithsonian Institution and the American Museum of Natural History.

Although Peale's dioramas and fossils were the prototypes of the modern natural history museum, his "new" museum idea was waylaid by two problems (Sellers, 1980). The first he correctly identified as the need for reinterpretation of the United States Constitution so that federal funds could be used to support collections of national importance. Indeed, it would not be too extreme to say that Peale hoped for a "first amendment" based upon collection needs. His Philadelphia Museum was used as a convenient repository for collections made on government expeditions to the West, but Peale was never given a budget for their curation (Dupree, 1957). The best of the Lewis and Clark collections went directly to Jefferson, who maintained them privately as part of his Poplar Forest estate. Peale, by contrast, received the problems—unwanted or damaged items (Jackson, 1978). Two rapidly growing and increasingly surly bear cubs from Jefferson is a case in point. After terrorizing Peale's family, with whom they shared cramped quarters in the State House basement, they were destroyed and mounted for display.

The second obstacle to the development of natural history in the United States was the location of type specimens. Type specimens are those specimens first given their scientific names and descriptions in press. Many type specimens of North American species (with notable exception of Peale's birds) were housed abroad. Botanical collections, for example, the Lewis and Clark herbarium, typically were given to individuals for description. Without stringent government controls, many of these valuable type materials were dispersed and lost.

The decline of Peale's museum after the founder's death in 1827 made real the need of a permanent home for government collections of natural history. Congress, however, did not give the management of federal collections serious attention until after 1842. The occasion was the return of the first purely scientific expedition, the U.S. Exploring Expedition under Lt. Charles Wilkes. As the expedition explored the

South Seas from 1838-42, former Secretary of War, Joel Poinsett organized a private receivership for collections (Kohlstedt, 1971). Poinsett (for whom the poinsettia flower is named) hoped the newly formed National Institution for the Promotion of Science could become a much needed national museum for natural history. The establishment of the Smithsonian Institution in 1846 (Goode, 1987), did not immediately help natural history. Joseph Henry was named director or head, emphasized pure research, rather than collections, and resisted investing the Smithsonian income (then $30,000 a year) in a museum building. Although he was the leading American scientist of his day, Henry's wishes fortunately did not prevail in an era of rapid expansion. After the Civil War, Americans settled 430 million acres west of the Mississippi River. The Territorial Surveys mapped new western regions, determined mineral resources, and assessed land use. By 1876, collections sent back by Lt. W. H. Emory's Military Reconnaissance (1846-47), the Gadsden Boundary Survey (1854-55), the Pacific Railroad Surveys (1853), and F.V. Hayden's Geological Surveys of the Territories (1869-78) necessitated a change of attitude (Goetzmann, 1966). After Spencer F. Baird succeeded Henry at the Smithsonian, Congress established the United States National Museum in 1879. At long last there was a new building for natural history collections (Goode, 1901).

During the interim, natural history was promulgated at the local level (see Bates, 1965). The Philadelphia Academy of Natural Sciences opened its collections to the public in 1828. By mid-century, the bird collections were the largest in the world. First curated in 1817 by Peale's youngest son Titian, the collections grew to 150,000 specimens. The more broadly based New-York Historical Society established a cabinet of natural history in 1816 with type material from the herbarium of Linnaeus. Eventually the members suppressed scientific directions in favor of history and art (Vail, 1954). New York City could also boast a Lyceum of Natural History, founded in 1818 on the model of the Academy of Natural Sciences. Despite the promise of its metropolitan location, the lyceum never gained the stature of its Philadelphia sister organization, and was not reorganized after fire destroyed its collections in 1866 (Fairchild, 1887). The New York Academy of Sciences now houses the lyceum's archives.

Impressed by Philadelphia's collections, Daniel Drake organized the Western Museum Society in the frontier city of Cincinnati (Shapiro and Zane, 1970). He hoped to promote natural history and local instruction in the Ohio Valley. Like Charles Willson Peale, Drake had vision. One of Drake's first preparators was the then unknown bird painter, John James Audubon. His new museum attracted the attention of notable western travelers including Titian Peale in 1819 (Poesch, 1961), but, by 1823, economic depression forced its shut-down.

Besides the initiative of enterprising individuals and scientific societies, various states initiated their own natural history collections. Following the inauguration of the New York State Natural History Survey in 1836, the survey's Cabinet of Natural History was located in Albany in 1843 (Merrill, 1920, followed by the Museum of Scientific and Practical Geology and General Natural History opened in 1870. The New York State Museum is the present beneficiary of these former survey headquarters. The city of Albany had enjoyed a surprisingly rich and diverse tradition of natural history. In 1809, the Henry Trowbridge Museum occupied the Old City Hall. Trowbridge purchased the New Haven Museum and other collections, and in 1821, designated his establishment as the State Museum, second only to Peale's museum. Fire damaged Trowbridge's operation in 1838, and in 1855, the press reported a novel solution: the collections were to form a floating museum on the Mississippi River. Because of the educational efforts of Amos Eaton and Stephen von Rensselaer, Albany displayed an unusual commitment to public natural history (Guralnick, 1975). James A. Hurst, a former taxidermist for the state survey, took advantage of the intellectual climate and opened his Albany Free Museum in 1871 (Hall, 1976), Faced with competition from the Great Eastern Menagerie, Murray's Great Railroad Circus and P. T. Barnum's Great Traveling Museum (which owned Peale's collections), Hurst was forced to close three years later. Of elaborate Victorian design, Hurst's displays were noteworthy for an early effort to exhibit human evolution: a stereoscopic slide of a "Gorilla from Africa, supposed to be our next of kin." In fact, the ape Hurst showed as a baboon; the gorilla was still a novelty, having been first described in 1847 in the *Boston Journal of Natural History.*

Like similar organizations in other cities, the Boston Society of Natural History struggled to maintain its identity through its collections (Kohlstedt, 1979). In 1860, the society received the first of a series of large gifts. A land grant followed from the state legislature that enabled the society to reopen to the public in 1864. The society's collections were small compared to the resources of nearby Harvard College, but the timing was right. Scientific lectures, institutes, and summer programs were vogue, and the new museum of the Boston Society of Natural History drew heavily for support on the local community of professors and students. Scientists F. W. Putnam and Alpheus Hyatt provided the society with some promise (Dexter, 1976; 1966). Their charismatic teacher, Louis Agassiz, however, was able to garner and divert a wider public than his circle of students (Lurie, 1960).

Like Peale, Agassiz advised his public to "read Nature," and like the energetic Philadelphian, he considered museum collections an indispensable asset to higher learning. In 1859, he coordinated considerable private, state, and university resources to found the Museum of Comparative Zoology at Harvard University. In a second grand effort in 1865, he organized the famous Thayer Expedition in Brazil to test Charles Darwin's revolutionary ideas about biological speciation. Despite the elaborate planning and carefully made collections, a complete report of the results, studies of the distribution of fish, was never published. Although his unilateral decision-making at the museum recently has been criticized, Agassiz was the first research biologist to run a permanent museum in the United States (Miller, et al., 1972). Agassiz thought of the museum as a storehouse, and through his network of contacts, he established synoptic collections for beginners and experts alike. Whereas Henry resisted the museum idea at the Smithsonian Institution, Agassiz embraced it. As Congress debated the proper mission of the Smithsonian, Agassiz brought under one roof scattered private collections and the holdings of defunct organizations. Eventually, these included those of the Boston Society of Natural History, which in turn included some of Peale's well-traveled artifacts and specimens once owned by members of the Philadelphia Academy.

Through two of its greatest students, Spencer Fullerton Baird and George Goode Brown, Agassiz's museum made an enormous contribution to the leadership of natural history in the United States. Under Baird's youthful supervision, the National Museum's collections, especially the birds, mushroomed. Baird, in turn, employed Goode, then a young ichthyologist, to arrange the Smithsonian and U. S. Fish Commission exhibits for the Philadelphia Centennial Exposition of 1876. The articulate Goode soon became the foremost American museum professional and an early historian of museum development (Goode, 1901). In charge of the National Museum after 1887 until his death in 1896, Goode expanded its purview to include all aspects of American culture, although the majority of objects are natural history specimens (Hellman, 1978). Agassiz's museum also fostered the talents of A. S. Bickmore and Henry A. Ward, founder of Ward's Natural Science Establishment in Rochester, New York (Kohlstedt, 1980). This cultivation of leadership within natural history continued in the 20th century, when the museum hired Frederick Lucas, James Chapin, and Ernst Mayr.

In 1869, Bickmore founded the American Museum of Natural History in New York City (Osborn, 1911). Bickmore wanted more outreach and learning opportunities for children. His idea gained loyal support from Theodore Roosevelt, Sr., Benjamin A. Field, Robert Colgate, William E. Dodge, and later J. Pierpont Morgan (Kennedy, 1968). The state of New York chartered the museum, which the commissioners of Central Park housed on the upper floors of the new park's Arsenal Building. In 1871, the museum wisely joined forces with the Metropolitan Museum of Art and, with the aid of Boss Tweed, secured the still-standing agreement with the city to provide museum buildings, maintenance, and security. The cornerstone for the American Museum's first building at its present location on Central Park West was laid in 1874. Seven years later, the multimillionaire banker, Morris K. Jesup, became museum president. Jesup attracted more funds, appointed scholarly curators, and personally financed expeditions, including Robert E. Peary's trek to the North Pole. Later, Jesup and his widow left six million dollars to the museum, a sum that has grown to 20 percent of the present endowment. The American Museums collections are second only to those of the National Museum (Hellman, 1969).

Nineteenth-century natural history involved the study of present organisms; it also involved knowledge of the past. Only nine dinosaurs were known before 1847. This number rose as fossil finds made by the Nebraska Survey were presented to the Academy of Natural Sciences. Intense competition among museums and their affluent patrons of the Gilded Age followed (Miller, 1970). In 1866, a new institution joined the fray, the Peabody Museum of Natural History, established at Yale University by George Peabody. After 1870, Othniel C. Marsh, Peabody's nephew and the museum's first director, conducted a series of expeditions which uncovered the first skeletons of *Brontosaurus, Stegosaurus, Diplodocus,* and *Triceratops.* These spectacular fossils were displayed with much fanfare. In addition to 80 new dinosaurs, Marsh described the first Cretaceous toothed birds, early marsupials, and his famous fossil horse series, including specimens of 30 American species. Through Marsh's efforts, the Peabody Museum's collection of Mesozoic mammals became the finest of its kind. The display of fossil horses was cited by both Charles Darwin and Thomas Henry Huxley to support the theory of evolution.

Not to be outdone, between 1881 and 1910, trustees of the American Museum of Natural History spent over $110,000 on 52 expeditions to western bone quarries. The materials from Bone Cabin, Wyoming, alone filled two railroad cars supplied by J. P. Morgan. In total, four major quarries yielded 493 specimens of Jurassic mammals. Furthermore, in 1895, the trustees used $35,000 from a secret fund of Morgan's to buy the collections of Edward Drinker Cope. Cope, a veteran of earlier government surveys led by F. V. Hayden in 1870–73 and 1857–79, was Marsh's bitter rival (Shor, 1874). He was also the nation's authority on fossil fish. A prodigious worker, Cope described 56 new dinosaurs, and the American Museum's splendid Hall of the Age of Mammals still displays 57 of his discoveries, including 28 type specimens. Many more still remain unpacked in store rooms.

Curator of the largest vertebrate paleontological collections in the world, Morgan's nephew, Henry Fairfield Osborn, capitalized upon the appeal of these fossil animals. As museum president from 1922 and 1930, he raised one million dollars and charged Walter Granger with the largest privately owned land expedition ever to leave the country, this time for the Gobi Desert. Osborn's methods worked. In one year alone, the much publicized find of a clutch of fossilized dinosaur eggs attracted contributions of $284,000. The dinosaur wars did not end in complete victory for New York. In 1896, the great industrialist and philanthropist, Andrew Carnegie, established his Museum of Natural History in Pittsburgh (McGinnnis, 1982). Impressed by Osborn's department of vertebrate paleontology, Carnegie ordered the museum's director, W. J Holland, to procure a dinosaur for $10,000, and to that end field collectors were sent west in 1898. Following the discovery of "Dippy," an almost compete ten-ton *Diplodocus*, A. S. Coggeshall designed a system for mounting dinosaurs that is still used today. Replicas of "Dippy" were sent to museums in Austria, Russia, Argentina, and Germany. In 1909, Earl Douglas found the unusually rich site at present-day Dinosaur National Monument, Utah, and the Carnegie Museum began to amass one of the world's outstanding collections of Mesozoic reptiles.

Exhibition techniques became increasingly important for natural history as "Dippy" demonstrated. In New York, Jesup supported the beautiful bird habitat groups at the American Museum developed by Frank M. Chapman for the present Hall of North American Birds, Birds of the World, and the Whitney Hall of Birds. Although Chapman had no college education, he rose from an assistant position in 1888 to head of the museum's ornithology department in 1908. Chapman, in turn, encouraged the fine bird-painter, Louis Agassiz Fuertes, and discovered the talented museum artist, Francis Lee Jaques. In Chicago, Marshall Field's gift of one million dollars and the success of the Columbian Exposition launched the Field Museum (Collier, 1969). Field hired the master preparator, Carl Akeley, to mount animals for the museum, including the pair of elephants in the Great Hall. In 1909, Akeley left the Field Museum to create dioramas for the American Museum. As museum president, Osborn emphasized the museum arts and interpretation, and, despite his controversial statements on race and eugenics, by 1827, he had given "museology," a term he coined, real application (Porter, 1983). From 1908 to 1932, he personally supervised the work of museum artists Charles Knight, R. Bruce Horsfall, and Erwin S. Cristman. He organized exclusive hunting clubs to bag exotic living species and finance the magnificent dioramas constructed under his presidency. Even at the height of the Great Depression, the museum's powerful trustees managed to meet the costs of these life groups, which ranged anywhere from $10,000 to $35,000 apiece. Exhibitions were becoming big business. Other museums imitated Osborn's approach to fund raising and organized affluent supporters to bag specimens for exhibitions. Thousands of animals were gathered in East Africa for the new Natural History Museum of Los Angeles County established in 1910. The L. A. museum received entire families of chimpanzees and monkeys, birds, insects, and plants from two wealthy enthusiasts, Maurice A. Machris and T. A. Knudsen. They traveled in a grand way, were accompanied in the field by 60 animal trackers, professional hunters, drivers, and cooks, as well as the museum's senior taxidermist, George Adams.

In recent years, the zealous collection policies of the turn of the century have been reevaluated as natural habitats all over the world are disturbed and destroyed (Squires, 1964; Parr, 1963). The sense of wonder and beauty that natural history inspires has traditionally motivated collections and justified the expense and consequences of their exhibition. In more recent times, policies of collection and exhibition have been shaped by considerations other than aesthetics or education. Ernst Mayr and Richard Goodwin (1956) were among the first to scrutinize biological materials in museum collections. Volume two of *The Preservation of Natural History Specimens* (1955–1968), edited by Reginald Wagstaffe and J. Havelock Fidler, and the 1968 symposium of the Biological Society of Washington, D.C., entitled "Natural History Collections, Past, Present, Future" were other early efforts to reexamine the role of collections in research. Since then, many others have discussed past and future problems faced by curators charged with preserving natural history materials in a changing social climate (Ripley, 1969; Evans, 1962). The antiquarian movement that revived natural history among 13th-century schoolmen in Europe has now a 20th-century counterpart. Will it continue to have profound impact on the understanding of natural history? As noted in the beginning of this paper, natural history is a museum word, and like artifacts, has come under museum care. Are museums really prepared for this responsibility? Certainly, directors and curators have become more self-conscious about research, collection, and exhibition decisions (Nicolson, 1974). As a result, museums of natural history in the United States have become the first to incorporate conservation of species into the fabric of their policies. Relying upon exhibits to convey their message, many new institutions will not have the luxury of their own systematic collections (Bloom *et al.*, 1984) In 1808, Alexander Willson urged readers of his *American Ornithology* to "examine better into the operations of nature [so that] our mistaken opinions, and groundless prejudices will be abandoned for more just, enlarged and humane modes of thinking." He was addressing the need to reevaluate the meaning of species. Perhaps museum visitors of the 21st century will be able to enjoy exhibits which better reflect these natural history ideals.

LITERATURE CITED

Bates, Ralph S. 1965. Scientific societies in the United States. 3rd ed. MIT Press, Cambridge, MA.

Bell, Whitfield J., Jr., Clifford K. Shipton, John C. Ewers, Louis Leonard Tucker, and Wilcomb E. Washburn. 1967. A cabinet of curiosities: five episodes in the evolution of American museums. University Press of Virginia, Charlottsville.

Best, Michael R. and Frank H. Brightman, eds. 1974. The book of secrets of Albertus Magnus. Oxford University Press, London.

Betts, John Rickards. 1959. P. T. Barnum and the popularization of natural history. J. History of Ideas, 20:353-368.

Blane, William Newnham. 1828. Travels through the United States and Canada. Baldwin, London.

Bloom, Joel N., Earl A. Powell III, Ellen Cochran Hicks, and Mary Ellen Munley, 1984. Museums for a New Century. American Association of Museums, Washington, D.C.

Bragg, Laura M. 1923. The birth of the museum idea in America. Charleston Mus. Quart., 1:3-13.

Collier, Donald. 1969. Chicago comes of age: the world's Columbian Exposition and the birth of the Field Museum. Field Museum Nat. Hist. Bull., 40:3-7.

Debus, Allen G. 1978. Man and nature in the Renaissance. Cambridge University Press, New York.

Dexter, Ralph W. 1966. Frederic Ward Purnam and the development of museums of natural history and anthropology in the United States. Curator, 9:151-155

———. 1976. Contributions of the Salem group of Agassiz zoologists to natural science education. BIOS, 47:25-30.

Dupree, A. Hunter. 1957. Science in the federal government a history of policies and activities to 1940. Belknap Press of Harvard University Press, Cambridge, MA.

Ellis, Richard P. 1966. The founding, history, and significance of Peale's Museum in Philadelphia, 1785-1841. Curator, 9:235-258.

Evans, J. W. 1962. Some observations, remarks, and suggestions concerning natural history museums. Curator, 5:77-93.

Fairchild, Herman LeRoy. 1887. A history of the New York Academy of Sciences, formerly the Lyceum of Natural History. Published by the author, New York.

Farber, Paul Lawrence. 1977. The development of taxidermy and the history of ornithology. Isis, 68:550-566.

Faxon, Walter. 1915. Relics of Peale's Museum. Bull. Mus. Comp. Zool., 59:117-148.

Flexner, James Thomas. 1954. The light of distant skies, 1760-1835. Harcourt, Brace, New York.

Goetzmann, William H. 1966. Exploration and empire; the explorer and the scientist in the winning of the American West. Alfred A. Knopf, New York.

Goode, George Brown, ed. 1897. The Smithsonian Institution 1846-1896: the history of its first half century. n.p., Washington, D.C. Reprint. 1980. Arno Press, New York.

————. 1889. Museum-History and Museums of History. Pap. Amer. Hist. Assoc., 3:495-579. Reprinted in Annual Report of the Board of Regents of the Smithsonian Institution. . . 1897. Report of the U.S. National Museum, Part 2. 1901. Government Printing Office, Washington, D.C.

————. 1901. The genesis of the U.S. National Museum. In Annual Report of the Board of Regents of the Smithsonian Institution. . . 1897. Report of the U.S. National Museum, Part 2. Government Printing Office, Washington, D.C.

Greene, John C. 1966. The Founding of Peale's Museum. In Bibliography & natural history (Thomas R. buckman, ed.) University of Kansas Libraries, Lawrence.

Guralnick, Stanley M. 1975. Science and the ante-bellum American College. American Philosophical Society, Philadelphia.

Hatt, Robert T. 1976. James A. Hurst, New York's first state taxidermist. NAHO, 9:3-17.

Hellman, Geoffrey Theodore. 1978. The Smithsonian: octopus on the mall. Lippincott, Philadelphia. 1967. Reprint. 1978. Greenwood Press, Westport, CT.

————. 1969. Bankers, bones & beetles; the first century of the American Museum of Natural History. Natural History Press, Garden City, N.Y.

Honour, Hugh. 1975. The new golden land: European images of America from the discoveries to the present time. Pantheon Books, New York.

Jackson, Donald D., ed. 1978. Letters of the Lewis and Clark expedition, with related documents, 1782-1854. 2nd ed. University of Illinois Press, Urbana.

————. 1981. Thomas Jefferson & the Stony Mountains: exploring the West from Monticello. University of Illinois Press, Urbana.

Kennedy, John Michael. 1968. Philanthropy and science in New York City: the American Museum of Natural History 1868-1968. Unpublished Ph.D. dissertation, Yale University.

Kohlstedt, Sally Gregory. 1971. A step toward scientific self-identity in the United States: the failure of the National Institute, 1844. Isis, 62:339-362.

————. 1979. From learned society to public museum: The Boston Society of Natural History. In The organization of knowledge in modern America,

1860-1920 (Alexandra Oleson and John Voss, eds.). Johns Hopkins University Press, Baltimore.

———. 1980. Henry A. Ward: the merchant naturalist and American Museum development. J. Soc. Biblio. Nat. Hist., 9:647-661.

Lurie, Edward. 1960. Louis Agassiz: a life in science. University of Chicago Press, Chicago.

Martin, Edwin Thomas. 1952. Thomas Jefferson: scientist. Henry Schuman, New York.

Mayr, Ernst, and Richard Goodwin. 1956. Biological materials, Part 1. Preserved materials and museum collections. Biology Council National Research Council Publication, No. 399. National Academy of Sciences-National Research Council, Washington, D.C.

McGinnis, Helen J. 1982. Carnegie's dinosaurs: a comprehensive guide to Dinosaur Hall at Carnie Museum of Natural History (Martina M. Jacombs and Ruth Anne Matinko, eds.). Board of Trustees, Carnegie Institute, Pittsburgh.

McLean, Antonia. 1972. Humanism and the rise of science in Tutor England. Neale Watson Academic Pub., Inc., New York.

Meisel, Max. 1924-1929. A bibliography of American natural history; the Pioneer Century, 1769-1865; the role played by the scientific societies; scientific journals; natural history museums and botanic gardens; state geological and natural history surveys; federal exploring expeditions in the rise and progress of American botany, geology, mineralogy, paleontology and zoology. 3 vols. Premier Publishing Company, Brooklyn.

Merrill, George P., ed. 1920. Contributions to a history of American state geological and natural history surveys. Smithsonian Institution, U. S. Nat. Mus. Bull., No. 109. Government Printing Office, Washington, D.C.

Miller, Howard S. 1970. Dollars for research; science and its patrons in nineteenth-century America. University of Washington Press, Seattle.

Miller, Lillian B., Frederick Voss, and Jeannette M. Hussey. 1972. The Lazzaroni: science and scientists in mid-nineteenth century America. Smithsonian Institution Press, Washington, D.C.

Nicholson, Thomas D. 1974. NYSAM Policy on the acquisition and disposition of collection materials. Curator, 17:5-9.

Osborn, Henry Fairfield. 1911. The American Museum of Natural History, its origin, its history, the growth of its departments to December 31, 1901. 2nd ed. Irving Press, New York.

———. 1963. Civilization and environment: a program for museums. Canad. Mus. Assoc. Bull., 14:1-6.

Peale, Charles Willson. 1800. Discourse introductory to a course of lectures on the science of nature; with original music, composed for, and sung on, the occasion. Zachariah Poulson Jr., Philadelphia.

Peale, Charles Willson, and Ambroise Marie François Joseph Palisot de Beauvois. 1796. A scientific and descriptive catalogue of Peale's Museum. Samuel H. Smith, Philadelphia.

Perry, John. 1976. P. T. Barnum's American Museum. Early American Life, 7:14-17, 56.

Phillips, Maurice E. 1953. The Academy of Natural Sciences of Philadelphia. Trans. Amer. Phil. Soc., n.s. 43:266-274.

Poesch, Jessie J. 1961. Titian Ramsay Peale, 1799-1885, and his journals of the Wilkes Expedition. Mem. Amer. Phil. Soc., vol. 52. American Philosophical Society, Philadelphia.

Porter, Charlotte M. 1983. The rise to Parnassus: Henry Fairfield Osborn and the Hall of the Age of Man. Mus. Stud. J., 1:26-34.

———. 1986. The eagle's nest: natural history and American ideas, 1812-1842. University of Alabama Press. University.

Ripley, S. Dillon. 1969. The sacred grove: essays on museums. Simon & Schuster, New York.

Sellers, Charles Coleman. 1947. Charles Willson Peale. 2 vols. Mem. Amer. Phil. Soc., vol. 23, pt. 1-2. American Philosophical Society, Philadelphia.

———. 1953. Peale's Museum. Trans. Amer. Phil. Soc., n.s. 43:253-259.

———. 1969. Charles Willson Peale. Charles Scribner's Sons, New York.

———. 1980. Peale's Museum and "The New Museum Idea." Proc. Amer. Phil. Soc., 124:25-34.

———. 1980. Mr. Peale's Museum: Charles Willson Peale and the first popular museum of natural science and art. W. W. Norton, New York.

Shapiro, Henry D., and Zane L. Miller, eds. 1970. Physician to the West: selected writings of Daniel Drake on science & society. University Press of Kentucky, Lexington.

Shor, Elizabeth Noble. 1974. The fossil feud between E.D. Cope and O.C. Marsh. Exposition Press, Hicksville, N.Y.

Simpson, George Gaylord. 1942. The beginnings of vertebrate paleontology in North America. Proc. Amer. Phil. Soc., 86:130-188.

Squires, Donald f. 1969. Schizophrenia: the plight of the natural history curator. Mus. News, 48:18-21.

Vail, Robert William Glenroiel. 1954. Knickerbocker birthday: a sesquicentennial history of the New York Historical Society, 1804-1954. New York Historical Society, New York.

Wagstaffe, Regianld, and J. Havelock Fidler, eds. 1955-1968. The preservation of natural history specimens. 2 vols. Philosophical Library, New York.

Whalley, Joyce Irene. 1982. Pliny the Elder: Historia Naturalis. Victoria and Albert Museum, London.

Wilson, Alexander. 1808. American ornithology: or the natural history of the birds of the United States. Vol. 1. Bradford & Inskeep, Philadelphia.

This paper has been expanded as a bibliographic essay in *Museums: A Reference Guide*. Michael S. Shapiro, ed. 1990. Greenwood Press, Westport CT.

Natural History Museums: Directions for Growth
Paisley S. Cato and Clyde Jones, editors
Texas Tech University Press, Lubbock, 1991, iv+252 pp.

ARE WE GOING IN CIRCLES?

David Lintz

Abstract.—Private collections of 16th- and 17th-century Europe formed the cabinets containing curiosities of natural and artificial objects for the owner and his friends. These gave rise to the early museums of the 18th century. By the mid-1800s many of these collections became public and modern museums were born. Early museums in Europe and America struggled with their scope and role as institutions for collection, preservation, and exhibition. Many concepts of natural history museums have been handed down and changes and trends have taken circular routes, falling in and out of favor with museum professionals. The old is continually giving way to the new—or is it? Concepts of exhibition in natural history museums likewise have been circuitous and systematic exhibits have given way to habitat groupings and then to didactic exhibits and back again. Today's natural history museums are redesigning their efforts to reflect the "modern" concept of man's relationship to the environment.

We are going to take a trip through time with vignettes presented for your thinking. These will be presented without much comment. Pay particular attention, during the several circles in which we may be traveling, to the area of museum exhibition and design. Although these are all opinions, they represent the thinking of the time by prominent museum professionals.

What are the things for which a man might want to strive? First, ". . . a most perfect and general library . . .;" second, ". . . a spacious and wonderful garden. . .;" third, "a goodly, huge cabinet, wherein whatsoever the hand of man by exquisite art or engine has made rare in stuff, form or motion; whatsoever singularly, chance, and the shuffle of things hath produced; whatsoever Nature has wrought in things that want life and may be kept; shall be sorted and included. The fourth such, a still-house, so furnished with mills, instruments, furnaces, and vessels as may be a palace fit for a philosopher's stone" (Impey, 1985). The year was 1594; those were the words of Sir Frances Bacon.

Bacon presented a neat, compact characterization of the nature of the museum in the late 16th century. In terms of function, little has changed. Museums are still in the business of keeping and sorting the products of man and nature and in promoting understanding of their significance. But what of museum exhibits?

Museums have been in the exhibit business for a long time, but emphasis and value have varied. Lately, exhibitions and public functions of museums have increased in importance and forced us to rethink our approaches to exhibits. Now we are beginning to develop new techniques for the exhibition of our collections for education and enjoyment. Or are we?

The most famous museum was founded by Ptolemy I about 300 BC in Alexandria. It was not, however, a museum in our sense of the word (Colbert, 1961). It was more similar to an academy or college of scholars with a collection of related material objects than to the modern concept of museum. (Leigh-Browne, 1968).

The museum, as we presently know it, had its beginnings during the Renaissance, when rich and noble men had the time, the means, and the intellectual curiosity to gather together objects of all kinds to form their personal collections (Colbert, 1961). These early European museums often were the private collections of the aristocracy and wealthy patrons of the arts and sciences. They were seldom open to the public and were used to show off to friends who had similar interests and level of knowledge. More often than not these holdings were in huge trophy rooms where everything was placed on view (Neal, 1987). In most all instances the collections were the exhibits. The furniture in the rooms was used for both display purposes and for storage (Eri, 1985).

From the beginnings of modern museum collections in the 15th and 16th centuries until the 19th century, objects were stored and exhibited together either in cabinets or on shelves. These arrangements reached a height of popularity during the period when the collections of the *Kunstkammer-* and *Wunderkammer-*type curio cabinets were taking shape and developing (Eri, 1985.)

These early exhibits generally were arranged as collections, with little or no order. There were some individuals, however, who thought differently. The principle that objects worth preserving should be divided into separate groups is enunciated in Samuel Quiccheberg's *Teatrum sapientiae*, published in 1565. It was a principle that applied to the collection of objects, the kind of furniture used for storing them, and the way in which they were arranged and displayed (Eri, 1985).

These cabinets of curiosities were not considered by all to be an essential part of the house of learning. Descartes disliked the whole business of curiosity. There was wavering and doubt in the academy also. In 1684, one scholar decried the fakes and misinterpretations, but in view of recent surprising discoveries, what should one not believe? For another authority, in 1714, cabinets enabled one to see exotica without traveling and therefore performed a useful service, but their exhibits were often unrepresentative or trivial fragments of nature that only wasted time (Impey, 1985).

Neickel's *Museographia*, published in 1727, gave a detailed description of the ideal exhibition room. In addition to giving the measurements, window placement, and color scheme, he gave the number, order, and positioning of display shelves. He stated that exhibits should be clearly displayed and appropriately placed according to size (Eri, 1985).

An epoch-making change occurred when museum exhibits were separated from the collections kept in storage. This separation of museum collections from display objects signalled the triumph of a new, modern, educational approach—one still valid today—to exhibiting museum objects. This approach emerged in the United States in the last decade of the 19th century and stemmed from a new concept of public education—a sense of responsibility to the public and a new approach to the purposes of museums (Eri, 1985).

Eventually, collections on exhibit were ordered and objects displayed in context with each other. In the last several years museums have attempted to use their objects to illustrate story lines and we have called them interpretive exhibits. Many times specimens are placed in their particular environment or habitat. Exhibits now combine a variety of audiovisual techniques, models, and artificially fabricated settings. Is this trend positive? Is it new? Are our techniques effective? (Peart, 1984). Some have said the emphasis on exhibit has gone too far, that we are close to the point where exhibits do little exhibiting of the museum's collections and hence exhibit very little of the curator's knowledge.

But, back to the vignettes. In 1683, a museum was first opened to the public by the Duke of York. The collections were the cabinet of curiosities gathered in the first half of the 17th century and given by Elias Ashmole to his university. Thus began the Ashmolean Museum of Art and Archaeology at the University of Oxford. Late in the 17th century, other museums began to open their doors to the general public, notably the Tower of London and the University Museum at

Basel. The British Museum, founded in 1753, also opened its doors for public visitation.

In America during this time, Peale's Museum in Philadelphia was at its zenith. For a privately owned museum, Peale's enjoyed an unusually long life (1784–1845), and it was the object of envy of European naturalists. Charles Willson Peale became a museum-keeper almost by accident. He was primarily an artist and kept portraits in his house for callers to see. In 1784, a local naturalist asked him to draw some "mammoth bones" that he had just acquired. Peale put them on view at his gallery as a special attraction. Because of the good response, Peale sought other natural curiosities and was presented with a paddle-fish. Friends donated specimens; Peale and his sons became ardent collectors. Sea captains sent back creatures from distant places. The museum grew. Specimens of birds were arranged in glass cases along a wall with the birds on a limb or rock, and with painted backgrounds showing foliage and terrain appropriate to the specimen—the first habitat groups. Specimens were classified according to Linnean arrangement. In all of his exhibits Peale tried to combine popular appeal and education with scientific accuracy (Richardson *et al.*, 1983).

Peale retired in 1810 at the age of 69. The management of the museum went to his son Rubens. In 1827, Rubens left to organize rival Peale's Museums in Baltimore and New York. The younger son, Titian, took over; the museum began its decline. He was knowledgeable in natural history and this led him to the field rather than to the cabinet, and he never involved himself in the affairs of the museum.

By the late 1820s and 1830s Peale's museum had several rivals. These included Rubens' other Peale's Museums, Scudder's American Museum in New York, and the Western Museum in Cincinnati. In 1841 Scudder's American Museum was bought by an unknown young man, Phineas T. Barnum, who in the next few years introduced new standards of scale, sensationalism, and vulgarity to the art of drawing crowds. He eventually swallowed up almost every private museum in America into his Grand Colossal Museum and Greatest Show on Earth. Peale's Museum could not compete in the Barnum manner, and by the 1840s, had lost whatever scientific importance it had once possessed. It was sold by the sheriff in 1845, and the collections widely dispersed. The demise of Peale's

Museum (and many other similar, privately owned museums in Britain and America) was inevitable. The concept of museums was changing and miscellaneous jumbles of "curiosities" were no longer good enough to attract the public.

In the early 1800s in London, Bullock presented his cultural material in simulations, offering traditional case displays with panoramic views and facsimiles around the gallery (Hall, 1987).

During the 19th century the function of exhibitions began to change because the common man no longer found museums so intimidating. Many of these museums were organized on rigid plans, with the cases on strict grids by taxonomy. The collections were organized to appeal to the connoisseur, collector, or scholar, and the displays were either arranged aesthetically, or by classfication or chronology (Hall, 1987).

In 1820, the Western Museum of Cincinnati was founded. It expired in 1867. For the first three years it operated as a center of science and was committed to the ideal of extending man's knowledge of the natural world. For the remaining 44 years of its existence it maintained a scientific facade, but functioned on the premise that a museum should entertain, enthrall, and frighten patrons rather than enlighten or educate them. Vulgarized and converted into a freak and horror show, the Western Museum became one of the best-known entertainment sites in the United States—the first Disneyland of the West (Whitehill, 1965). Its demise was due to the unyielding reality of operating expenses.

The year was 1881. George Brown Goode was Assistant Director and Executive Officer of the United States National Museum. In his first report he outlined a new kind of National Museum with a "perfect plan of organization and a philosophical system of classification" with three aspects. One is a Museum of Record—a treasure house going back to Ptolemy's Mouseion in Alexandria, to the Greek temples and medieval churches, and to the European royal collections. Next, there is a Museum of Research that will house the collections gathered for taxonomic studies and other research. Finally, there will be the Educational Museum devoted to the "systematic exhibition of industry for the instruction of visitors, the improvement of public taste, and the fostering of the arts of design." Goode thought that the kind of museum he visualized had not yet been achieved anywhere. It will "show, arranged according to one consistent plan, the resources of the

earth and the results of human activity in every direction"
(Alexander, 1983).

In the field of exhibition, Goode stated "the people's
museum should be much more than a house full of specimens
in glass cases. It should be a house full of ideas, arranged with
strictest attention to system" (Alexander, 1983). Goode believed
in exhibiting a series of collections in the main halls of the National Museum for laymen and the general public, and exhibiting a study or reserve collection in laboratories and storerooms
for research use by specialists and students. This dual concept
was also becoming popular in Europe, and when the British
Museum (Natural History) moved to South Kensington, about
1880, its collections were soon placed on this plan. Sir William
Henry Fowler, its director, referred to this plan as the "New
Museum Idea" (Alexander, 1983).

The year was 1939. Laurence Vail Coleman (1939) wrote
that "it is customary to speak of exhibits as though there were
two distinct kinds—didactic and aesthetic." These elements
enter in different proportions into exhibits. Coleman also
wrote ". . .the perfecting of exhibits to convey biological information—is already old, going back to Agassiz [Museum of
Comparative Zoology, 1852] whose arrangements of animals
to show their classes and distribution were the earliest of
'idea' exhibits. It is easy to overlook that advance, which
marked a departure from the unorganized displays of early
museums, because now we are used to more involved exhibits
designed to bring out the structure, functions, development,
and habits of living things. . . ." Similarly, but on a larger
scale, he continued, "we are accustomed to complex associations of exhibits such as are found in a hall of ocean life, an
African hall, or rooms devoted to theoretic subjects such as
evolution and psychology."

"There is a growing tendency to organize exhibits around . . .
key installations, breaking up the story into chapters and
featuring a key point in each chapter. Here aesthetic display
and didactic display come together, for the feature exhibit is
very likely to be a work of art" (Coleman, 1939).

Coleman (1939) also declared, "Science exhibits, for the
last half-century, have been dominated by the habitat group. . .."
This became possible because museum curators were turning
from description to the study of habits, and were ready—
despite the resistance of conservatives among them—to give

the public an improvement upon the accustomed rows of specimens arranged in scientific classification and 'assorted according to size. Habitat groups are being challenged as things of the past. That seems to go too far; but groups have certainly been overdone and are likely soon to take their proper place in the scheme of things" (Coleman, 1939).

"Miniature groups, or dioramas—reduced in scale, and therefore relying upon models instead of mounted specimens— have a much larger field of usefulness than they have yet found. Mammal groups with small colored figures . . . are sure to be well accepted after a time" (Coleman, 1939).

The year was 1939. Albert E. Parr explained that the natural history museums of today can no longer justify themselves solely on the existence of their collections. The museum must no longer devote its entire effort to satisfying the abstract scientific curiosity of a comparatively small group of enthusiasts. Instead, increasing emphasis must be placed on the more concrete problems in man's relation to nature which are of common concern to all mankind (Parr, 1959).

Parr (1959) also stated "That the availablity of the traditional exhibits, artifically segregated by subject, is of great importance to formal instruction is undeniable, but whether they constitute the most significant and valuable contribution the museum can make to systematized education is questionable. The most valuable contribution the museum can make to further the true goals of all formal education is probably found in an . . . attempt to create a picture and philosophy of nature as a whole. . . "

". . . Every visitor would be . . . exposed to the educational effects of a continuous arrangement of exhibits, telling a balanced and coherent story of nature as an entity in itself and as the environment of man. . ."(Parr, 1959).

The year was 1951. The new Hall of Evolving Life at the Los Angeles County Museum was about to open. Chester Stock wrote, ". . . a hall is now undergoing modification to develop a plan to utilize the display area to express an entirely different exhibit concept. Thus, the old is giving way to the new. . . . In other words, rather than emphasing the state of existence by a display of specific facts, the present demonstration attempts to emphasize the nature of change in living organisms and some of the reason for that change" (Stock, 1951).

Also, in 1951, at the opening of the American Museum's Warburg Hall, the author of an article in *Museum News* wrote "the hall is the first unit to be completed under the museum's broad plan for exhibits to analyze relations of man and nature, emphasing the study of man's environment as a practical necessity. [The exhibits will] . . . explain the area's geological history, soil, climate, vegetation, animal life, and the effects of settlement." A. E. Parr, spoke of the museum's new methods of presentation: "In this new hall we deal with nature as the environment of man, subjected to his good or bad influences. We take the totality of nature rather than any particular type of life as our main theme. We have attempted to relate our story to the daily, or at least weekend, life of our visitors" (The American Association of Museums, 1951).

The year was 1952. Karl Schmidt stated that "by means of the museum, we enable ourselves to touch reality, and to go back and touch reality again and again. However much the university museum may engage in exhibition, it is difficult to envisage in it the unthinking overemphasis of the habitat group, which has been so much a feature of the American public museum."

The year was 1958. Lothar Witteborg of the American Museum of Natural History said "The primary function of a museum and its exhibits is to educate. To achieve this end at a natural history museum, exhibits should be planned in which actual life is illustrated and in which native skills and cultures are displayed. Nothing should be shown merely because it is ancient or has curiosity value. The natural history museum should take elements from nature and from life itself along with theories, concepts, and philosophies achieved through scientific research, and combine them all into a meaningful presentation which tells a story" (Witteborg, 1958).

The year was 1961. Although many other kinds of displays declined in the shadow of the habitat group, it was primarily the systematic exhibits that suffered a direct loss of museum space. Other subjects were eclipsed, but systematics were being replaced. "To add insult to injury, a specimen removed from a systematic array will often turn up again in the enemy's camp as a slightly remodeled member of a happy habitat group. The process is still continuing, and in small museums it can actually become a quite destructive influence, when a brief, but sound and, in its fashion, comprehensive systematic digest of nature is set aside to give room for a few space-consuming glimpses of special and spectacular situations that are insufficient to develop any continuity of theme or integrated coverage of subject" (Parr, 1961).

The year was 1963. Albert E. Parr (1963) wrote "A relaxation of the rules of admission for new exhibits is already evident. It used to be that the natural history museums would not permit themselves to show any details of natural events and environments that had not been verified by direct observation, and that the entire situation portrayed in the exhibit would have to illustrate a normal and usual condition of life for the species involved. Exhibits based only upon more or less well-informed and intelligent surmise are infiltrating our halls, and the commonly characteristic is often set aside in favor of the uncommonly spectacular but actually unimportant actions and circumstances. Perhaps some day soon the absence of positive disproof will take the place of proof positive as justification for our public spectacles."

The year was 1969. The current Associate Director at Carnegie wrote ". . . in our anxiety to develop the best methods to serve most of the general public, we've lost our sense of proportion. The general public is that class of our audience that is for the most part neither well enough informed nor sufficiently motivated to reap great benefits from observation and study of objects and ideas presented without considerable guidance. In consequence we devote more and more space to public-oriented, dramatic, 'teaching' exhibits, and most of the areas in our buildings not occupied by essential storage rooms, libraries, laboratories, exhibit shops, and service areas is or is in the process of being given over to this legitimate, useful, successful, and popular sort of display. Unfortunately, we became so convinced over the last thirty years that we should expand our exhibit service to the general public and we worked so hard to achieve this desired goal that, shamefully, we began to neglect to numerically inferior but equally important and vital pre-informed, pre-motivated, purposeful, serious, usually lone, knowledge-seekers—the informed layman. Informed laymen need the open storage we discarded—subject by subject exhibits. . . ." He suggested that we think very seriously about our need to be old fashioned (Swauger, 1969).

The year was 1972. A designer lamented that the end results of our exhibits are often a three-dimensional textbook

or a magically-lit diorama, and that we are relying on standard, timeworn approaches to exhibits and displays. He
pleaded for exhibits in which the visitor can participate—
learning through doing (Morano, 1972). This was something,
by the way, that Peale had done in his museum 150 years earlier.

The year was 1973. According to Shettel (1973), the didactic exhibit is the most important type of museum exhibit because it makes available for visitors the opportunity to
"increase their knowledge and/or to change their beliefs and
attitudes toward a wide variety of things not otherwise available." He defined didactic exhibits as those "which appear to
have an instructional or educational role to play. These exhibits tell a story, explain a process, define a scientific principle." He contended that the main museum effort should be
channeled into the production of didactic exhibits that
reflect the highest state of the art.

The year was 1981. A new exhibit has opened at the Royal
British Columbia Museum. "Living Land Living Sea" is a large
natural history exhibit that has open dioramas of forest environments and seashore habitats, closed dioramas of Ice Age
mammals, river habitats, and small didactic exhibits—a total
of 46 exhibits. What is the visitor response to the key exhibit
type in the gallery—the diorama? A survey developed and administered by Peart and Kool (1988) gave some answers.
Knowledge gain did occur, but the majority of intended mes-
sages were not communicated. Attitudinal change did not
occur. The attracting and holding power of the gallery was
low. Visitor flow did not follow the designed pattern, nor did
visitors stay in the gallery for the intended time period. Key
exhibits did not receive the necessary attention, but visitors
were attracted to and held at certain exhibits for other than
the intended purpose. There was, however, an overwhelming
positive enjoyment reaction.

Is the gallery a success? Are dioramas the answer? The
visitor might vote yes, the museologist, no. The exhibits
judged to be the most successful in behavioral terms (the
larger "concrete" exhibits) were not the most successful in
educational terms. The diorama exhibits are not the best
ones for communicating ideas. They may "wow" visitors, but if
learning is to occur, the small exhibit whose message can be
obtained in a relatively short time is the best choice.

The year is 1988. Are we going in circles? You be the judge. A designer recently observed, and rightly so, ". . . that both designer and curator today are very much within a tradition and they must recognize how hard it is to find completely new forms of display" (Hall, 1987).

LITERATURE CITED

Alexander, E. P. 1983. Museum masters. Amer. Assoc. State Local Hist., Nashville, x+428 pp.

American Association of Museums. 1951. American Museum opens new Warburg Memorial Hall. Museum News, 29(4)1-8.

Colbert, E. M. 1961. What is a museum? Curator, 4:138-146.

Coleman, L. V. 1939. The museum in America: a critical study. Amer. Assoc. Mus., Washington, 3 volumes, 730 pp.

Ellis, R. P. 1966. The founding, history, and significance of Peale's museum in Philadelphia, 1785-1841. Curator, 9:235-258.

Eri, I. 1985. A brief history of the show-case. Museum, 37:71-74.

Hall, M. 1987. On display: a design grammer for museum exhibitions. Lund Humphries, London, 256 pp.

Impey, O., and A. MacGregor, eds. 1985. The origins of museums. Clarendon Press, Oxford, xiii + 335 pp.

Leigh-Browne, F. S. 1968. Preface in Great museums of the world—British Museum London. Newsweek, Inc., New York.

Morano, V. J. 1972. Something old under the sun. Curator, 15:131-138.

Neal, A. 1987. Help for the small museum. 2nd Ed. Pruett Publishing Co., Boulder, xi+176 pp.

Parr, A. E. 1959. Mostly about museums from the papers of A. E. Parr. Amer. Mus. Nat. Hist., New York, 112 pp.

———. 1961. The revival of systematic exhibits. Curator, 4:117-137.

———. 1963. The functions of museums: research centers or show places. Curator, 6:20-31.

Peart, B. 1984. Impact of exhibit type on knowledge gain, attitudes, and behavior. Curator, 27:220-237.

Peart, B., and R. Kool. 1988. Analysis of a natural history exhibit: are dioramas the answer? Inter. Jour. Mus. Manag. Curatorship, 7: 117-128.

Richardson, E. P., B. Hindle, and L. B. Miller. 1983. Charles Willson Peale and his world. Harry N. Abrams, Inc., New York, 272 pp.

Schmidt, K. P. 1952. The function of the university museum. Museum News, 30(2):5.

Shettel, H. H. 1973. Exhibits: art form or educational medium. Museums News, 52(1):32-41.

Stock, C. 1951. New Hall of Evolving Life at Los Angeles County Museum. Museum News, 29(1):6.

Swauger, J. L. 1969. Topless girl guides or we have a need to be old-fashioned. Curator, 12:307-318.

The American Association of Museums. 1951. American Museum opens new Warburg Memorial Hall. Museum News, 29(4):1-8.

Whitehill, W. M. 1965. History of museums in the United States. Curator, 8:5-54.

Witteborg, L. P. 1958. Design standards in museum exhibits. Curator, 1:29-41.

CONTRIBUTORS

ALLEN S. BOHNERT Regional Curator, Rocky Mountain Region, National Park Service, Box 25287, Denver, CO 80225

PAISLEY S. CATO Department of Wildlife and Fisheries Science, Texas A&M University, College Station, TX 77843 *Present address:* Virginia Museum of Natural History, 1001 Douglas Avenue, Martinsville, VA 24112

JERRY R. CHOATE Fort Hays State Museums, Fort Hays State University, Hays, KS 67601

JUDY DACUS Educational Research Center, New Mexico State University, Las Cruces, NM 88003

JANE E. DEISLER-SENO Corpus Christi Museum, 1900 N. Chaparral, Corpus Christi, TX 78401

LOUISE LAURETANO DEMARS Peabody Museum of Natural History, Yale University, P.O. Box 6666, 170 Whitney Avenue, New Haven, CT 06511-8161

P. LYNN DENTON Texas Memorial Museum, 2400 Trinity Street, Austin, TX 78705

AMY LYN EDWARDS The University of Georgia, Museum of Natural History, Athens, GA 30602

JEFFRY GOTTFRIED Oregon Museum of Science and Industry, 4015 SW Canyon Road, Portland, OR 97221

PHILIP S. HUMPHREY Museum of Natural History, University of Kansas, Lawrence, KS 66045

CLYDE JONES Department of Biological Sciences and The Museum, Texas Tech University, Lubbock, TX 79409

JOSHUA LAERM Museum of Natural History, University of Georgia, Department of Zoology, Athens, GA 30602

DAVID LINTZ Strecker Museum, Baylor University, Waco, TX 76798

ELIZABETH PATTON Office of Public Education, Museum of Natural History, University of Kansas, Lawrence, KS 66045

CHARLOTTE M. PORTER Florida Museum of Natural History, University of Florida, Gainesville, FL 32611

JUDITH READER Division of Instructional Resources, Corpus Christi Independent School District, P.O. Box 110, Corpus Christi, TX 78403

CAROLYN L. ROSE Department of Anthropology, National Museum of Natural History, Smithsonian Institution, Washington, D.C. 20560

DAVID J. SCHMIDLY Department of Wildlife and Fisheries Sciences, Texas A&M University, College Station, TX 77843

SALLY Y. SHELTON Texas Memorial Museum, Materials Conservation Laboratory, BRC 122, 10100 Burnet Road, Austin, TX 78758

CATHERINE CARTER SHROPSHIRE Mississippi Museum of Natural Science, 111 North Jefferson St., Jackson, MS 39202

R. THOMAS SHROPSHIRE Mississippi Museum of Natural Science, 111 North Jefferson St., Jackson, MS 39202

VALEEN SILVY Rt. 4, Box 317AA, College Station, TX 77840

JOHN E. SIMMONS Museum of Natural History, University of Kansas, Lawrence, KS 66045

REBECCA SMITH Education Department, New Mexico Museum of Natural Science, P.O. Box 7010, Albuquerque, NM 87194-7010

MARGO SUROVIK-BOHNERT 10240 W. 34th Avenue, Wheat Ridge, CO 80033

PETER B. TIRRELL Oklahoma Museum of Natural History, The University of Oklahoma, Norman, OK 73019